Lecture Notes in Statistics

Vol. 1: R. A. Fisher: An Appreciation. Edited by S. E. Fienberg and D. V. Hinkley. XI, 208 pages, 1980.

Vol. 2: Mathematical Statistics and Probability Theory. Proceedings 1978. Edited by W. Klonecki, A. Kozek, and J. Rosiński. XXIV, 373 pages, 1980.

Vol. 3: B. D. Spencer, Benefit-Cost Analysis of Data Used to Allocate Funds. VIII, 296 pages, 1980.

Vol. 4: E. A. van Doorn, Stochastic Monotonicity and Queueing Applications of Birth-Death Processes. VI, 118 pages, 1981.

Vol. 5: T. Rolski, Stationary Random Processes Associated with Point Processes. VI, 139 pages, 1981.

Vol. 6: S. S. Gupta and D.-Y. Huang, Multiple Statistical Decision Theory: Recent Developments. VIII, 104 pages, 1981.

Vol. 7: M. Akahira and K. Takeuchi, Asymptotic Efficiency of Statistical Estimators. VIII, 242 pages, 1981.

Vol. 8: The First Pannonian Symposium on Mathematical Statistics. Edited by P. Révész, L. Schmetterer, and V. M. Zolotarev. VI, 308 pages, 1981.

Vol. 9: B. Jørgensen, Statistical Properties of the Generalized Inverse Gaussian Distribution. VI, 188 pages, 1981.

Vol. 10: A. A. McIntosh, Fitting Linear Models: An Application on Conjugate Gradient Algorithms. VI, 200 pages, 1982.

Vol. 11: D. F. Nicholls and B. G. Quinn, Random Coefficient Autoregressive Models: An Introduction. V, 154 pages, 1982.

Vol. 12: M. Jacobsen, Statistical Analysis of Counting Processes. VII, 226 pages, 1982.

Vol. 13: J. Pfanzagl (with the assistance of W. Wefelmeyer), Contributions to a General Asymptotic Statistical Theory. VII, 315 pages, 1982.

Vol. 14: GLIM 82: Proceedings of the International Conference on Generalised Linear Models. Edited by R. Gilchrist. V, 188 pages, 1982.

Vol. 15: K. R. W. Brewer and M. Hanif, Sampling with Unequal Probabilities. IX, 164 pages, 1983.

Vol. 16: Specifying Statistical Models: From Parametric to Non-Parametric, Using Bayesian or Non-Bayesian Approaches. Edited by J. P. Florens, M. Mouchart, J. P. Raoult, L. Simar, and A. F. M. Smith. XI, 204 pages, 1983.

Vol. 17: I. V. Basawa and D. J. Scott, Asymptotic Optimal Inference for Non-Ergodic Models. IX, 170 pages, 1983.

Vol. 18: W. Britton, Conjugate Duality and the Exponential Fourier Spectrum. V, 226 pages, 1983.

Vol. 19: L. Fernholz, von Mises Calculus For Statistical Functionals. VIII, 124 pages, 1983.

Vol. 20: Mathematical Learning Models – Theory and Algorithms: Proceedings of a Conference. Edited by U. Herkenrath, D. Kalin, W. Vogel. XIV, 226 pages, 1983.

Vol. 21: H. Tong, Threshold Models in Non-linear Time Series Analysis. X, 323 pages, 1983.

Vol. 22: S. Johansen, Functional Relations, Random Coefficients and Nonlinear Regression with Application to Kinetic Data. VIII, 126 pages. 1984.

Vol. 23: D. G. Saphire, Estimation of Victimization Prevalence Using Data from the National Crime Survey. V, 165 pages. 1984.

Vol. 24: T. S. Rao, M. M. Gabr, An Introduction to Bispectral Analysis and Bilinear Time Series Models. VIII, 280 pages, 1984.

Vol. 25: Time Series Analysis of Irregularly Observed Data. Proceedings, 1983. Edited by E. Parzen. VII, 363 pages, 1984.

Lecture Notes in Statistics

Edited by D. Brillinger, S. Fienberg, J. Gani, J. Hartigan, and K. Krickeberg

27

Arnold Janssen
Hartmut Milbrodt
Helmut Strasser

Infinitely Divisible Statistical Experiments

Springer-Verlag
Berlin Heidelberg New York Tokyo

Authors

Arnold Janssen
Universität-Gesamthochschule-Siegen, Fachbereich 6 – Mathematik
Hölderlinstr. 3, 5900 Siegen, Federal Republic of Germany

Hartmut Milbrodt
Helmut Strasser
Mathematik VII der Universität Bayreuth
Postfach 3008, 8580 Bayreuth, Federal Republic of Germany

AMS Subject Classification (1980): 62A99, 62F99

ISBN 978-0-387-96055-5 ISBN 978-1-4615-7261-9 (eBook)
DOI 10.1007/978-1-4615-7261-9

Library of Congress Cataloging in Publication Data. Janssen, Arnold. Infinitely divisible statistical experiments. (Lecture notes in Statistics; 27) Bibliography: p. Includes indexes. "AMS subject classification (1980) 62A99, 62F99"–T.p. verso. 1. Mathematical statistics. 2. Statistical decision. I. Milbrodt, Hartmut. II. Strasser, Helmut, Dr. phil. III. Title. IV. Series: Lecture notes in statistics (Springer-Verlag); no. 27. QA276.J36 1985 519.5 85-4662

2147/3140-543210

PREFACE

This book is intended to give an account of the theory of infinitely divisible statistical experiments which started from LeCam, 1974. It includes a presentation of LeCam's basic results as well as new developments in the field.

The book consists of four chapters written by different authors. Chapters I, III and IV have been prepared in Bayreuth with the support of the Deutsche Forschungsgemeinschaft (DFG); Chapter II is part of its author's Habilitationsschrift, 1982 (Dortmund). For the reader's convenience, the chapters have been unified in presentation, without neglecting differences in the individual styles of writing.

The authors are grateful to Dr. C. Becker for carefully reviewing the manuscript. Furthermore, acknowledgements are gratefully extended to the DFG for partly subsidizing Dr. Becker and the second author by a grant. Some special words of thanks are due to Mrs. Witzigmann, who typed the final manuscript and its predecessors with patience and skill.

Universität Bayreuth und
Universität Dortmund,

Dezember 1984

A. Janssen

H. Milbrodt

H. Strasser

CONTENTS

Preface

Introduction 1

I. Limits of Triangular Arrays of Experiments 14
(H. Milbrodt and H. Strasser)

1. Basic Concepts 14
2. Gaussian Experiments 19
3. Introduction to Poisson Experiments 25
4. Convergence of Poisson Experiments 32
5. Convergence of Triangular Arrays 38
6. Identification of Limit Experiments 47

II. The Lévy-Khintchine Formula for Infinitely Divisible Experiments 55
(A. Janssen)

7. Preliminaries 55
8. Infinitely Divisible Probability Measures 58
9. The Lévy-Khintchine Formula for Standard Measures 63
10. The Lévy-Khintchine Formula for Arbitrary Regular Infinitely Divisible Statistical Experiments 89

III. Representation of Poisson Experiments 106
(H. Milbrodt)

11. Generalized Poisson Processes 106
12. Standard Poisson Experiments 111

IV. Statistical Experiments with Independent Increments 124

(H. Strasser)

13. Preliminaries 124
14. Experiments with Independent Increments 126
15. Existence and Construction of Experiments with Independent Increments 130
16. Infinitely Divisible Experiments with Independent Increments 133
17. Weak Convergence of Triangular Arrays to Experiments with Independent Increments 140
18. The Likelihood Process 144
19. Application to Densities with Jumps 149

Bibliography 154

List of Symbols 157

Author Index 160

Subject Index 161

INTRODUCTION

Over the last 30 years the asymptotic theory of statistical methodology has gone through a rapid development. First general results were obtained around 1950, notably by Wald and LeCam. Their fundamental papers then mainly relied on mathematical tools from classical analysis and on probabilistic limit theorems, the theory of which had been extensively developed up to then. Clearly, the structure of asymptotic statistics at that time was strongly influenced by this framework. Since then LeCam and others succeeded in exhibiting the mathematical structure underlying this theory by getting rid of unnecessary analytical regularity conditions. On the one hand, this development made it possible to put the classical asymptotic results of statistics into their present, fairly final form (cf. LeCam, 1972 and 1979). On the other hand, it provided tools, which may serve for the solution of problems that have been inaccessible so far.

The text in hand aims at presenting ideas which, in the authors' opinion, will turn out to be of importance for future developments. These ideas form a statistical analogue of the probabilistic theory presented e.g. in Gnedenko's and Kolmogorov's famous book "Grenzverteilungen von Summen unabhängiger Zufallsgrößen". Roughly speaking, we deal with limits of products of independent experiments. For this problem, LeCam developed a theory the structure of which is similar to the set-up of Gnedenko and Kolmogorov. The particular case of binary experiments is treated in LeCam, 1969, whereas the case of an arbitrary parameter space is dealt with in LeCam, 1974. LeCam's presentation already contains all basic ideas. But - since the proofs are sometimes incomplete - it requires careful elaboration. Among other things, the present volume aims at rendering a fairly complete and self-contained exposition of the general case.

In the first part we give an elaboration of LeCam, 1974, Chapter 8, trying to fill in those omissions, which a careful reader of LeCam's text may regard as technical gaps. Within the framework of probability theory the results of Gnedenko and Kolmogorov have been successfully extended to general groups (cf. Heyer, 1977). This aspect forms the basis of the second part of the present text where the problem of Lévy-Khintchine formulae for statistical experiments is tackled by group theoretic methods. LeCam's original paper contains implicitely an important but unproven assertion concerning the representation of so-called Poisson experiments. Relying on tools developed up to then a complete proof is given in part three. The last part illustrates the power of the general theory exposed so far. Here the theory is applied to generalize some results of Ibragimov and Has'minskii, 1972 and 1981, concerning the asymptotic behaviour of nonregular statistical experiments.

Now, we shall give a brief introduction into some of the basic ideas of asymptotic statistics, presenting a point of view which leads to the problems dealt with in the text. For simplicity, we restrict ourselves to the i.i.d. case.

The object of statistics is the problem of measurement in case of random experiments. The aim is the determination of a characteristic value, a deterministic "parameter", of an experimental arrangement which cannot be observed directly, since it is superimposed by a random perturbation. In the simplest case, this superposition is additive. If $\theta \in \mathbb{R}^1$ is the parameter, the observation X is

$$X = \theta + U$$

where the random variable U describes the random perturbation. If the distribution of U has a Lebesgue density h, then the distribution P_θ of X is of the form

$$P_\theta(B) = \int_B h(\cdot - \theta)d\lambda^1 , \quad B \in \mathcal{B}^1 .$$

Hence, the mathematical model of the random experiment is given by $(\mathbb{R}^1, \mathcal{B}^1, \{P_\theta: \theta \in \mathbb{R}^1\})$. Such a triplet $E = (\Omega, \mathcal{A}, \{P_\theta: \theta \in \Theta\})$ consisting of a family of probability measures $\{P_\theta: \theta \in \Theta\}$ on a measurable space $(\Omega, \mathcal{A})$ is called a statistical experiment. Experiments of the particularly simple form described above are called shift experiments.

Intuitively, it is clear that a single observation of X does not admit a reasonable inference as to the unknown parameter θ. Hence, the experiment E is independently replicated n-times, thus obtaining a new experiment

$$E^n := (\Omega^n, \mathcal{A}^n, \{P_\theta^n: \theta \in \Theta\}) .$$

In the case of shift experiments the i-th observation X_i may also be of the form

$$X_i = c_i\theta + U_i$$

where the scalar c_i describes the varying experimental conditions and the perturbations U_i are independent ($1 \leq i \leq n$). Generally speaking, if the experiment $E_i = (\Omega_i, A_i, \{P_{i\theta}: \theta \in \Theta\})$ corresponds to the i-th observation ($1 \leq i \leq n$), then the experiment pertaining to the independent performance of all of these experiments is

$$\bigotimes_{i=1}^{n} E_i := \left(\prod_{i=1}^{n} \Omega_i, \bigotimes_{i=1}^{n} A_i, \{ \bigotimes_{i=1}^{n} P_{i\theta}: \theta \in \Theta\}\right).$$

The statistician analyzing an experiment looks for reasonable decision functions which should be optimal in some sense. But a direct solution of such optimization problems is only possible for special types of experiments. Many practically important models remain untractable from that viewpoint. A promising way out of this dilemma is offered by the probabilistic statement that stochastic phenomena originating from the superposition of many independent components can often be approximated by much simpler models. This is also the basic idea within asymptotic statistics. Let us see what this approach leads to.

For this, let $E = (\Omega, A, \{P_\theta: \theta \in \Theta\})$ be a statistical experiment with pairwise unequal probability measures. The n-fold independent replication of E yields $E^n = (\Omega^n, A^n, \{P_\theta^n: \theta \in \Theta\})$, $n \in \mathbb{N}$. Letting $n \to \infty$, the sequence $(E^n)_{n \in \mathbb{N}}$ would "tend" to $E^{\mathbb{N}} := (\Omega^{\mathbb{N}}, A^{\mathbb{N}}, \{P_\theta^{\mathbb{N}}: \theta \in \Theta\})$. From a theorem of Kakutani (cf. Hewitt and Stromberg, 1969, Theorem 22.36) it follows that the probability measures $P_\theta^{\mathbb{N}}$, $\theta \in \Theta$, are pairwise orthogonal. Hence, the limit experiment $E^{\mathbb{N}}$ is deterministic, and cannot serve as a good approximate model for the E^n, $n \in \mathbb{N}$. Nevertheless, from the way (E^n) converges to $E^{\mathbb{N}}$ it is still possible to draw conclusions as to the quality of certain sequences of decision functions. This approach was chosen by Bahadur. However, the first works of Wald, 1943, and LeCam, 1953, already pointed into another direction, now known as "local approximation".

In order to explain the concept of local approximation we have to go back a little bit. A first idea can be obtained from the comparison of the law of large numbers and the central limit theorem. The law of large numbers asserts convergence in distribution of the arithmetic mean to a point mass. Of course, this measure is useless as an approximation of the distribution of the mean. The central limit theorem improves the law of large numbers in that it uses a sequence of scale factors to force convergence to a non-degenerate limiting distribution, which may be looked upon as a good approximation to the exact law of the mean. Similarly, the method of local approximation uses a rescaling of the original sequence of experiments to obtain non-deterministic limit experiments. These are usually accepted as a first approximation of the single experiments.

Before describing the rescaling procedure in detail, we want to remind the reader of the problem of equivalence of statistical experiments. At an early stage, the observation that different experiments may be equivalent from a statistical point of view led to the concept of sufficiency, the applicability of which is restricted to very particular situations. Extending ideas of Blackwell, 1951 and 1953, for finite experiments, LeCam developed a meaningful concept of equivalence of arbitrary statistical experiments. Intuitively speaking, two experiments $(\Omega_1, A_1, \{P_\theta: \theta \in \Theta\})$ and $(\Omega_2, A_2, \{Q_\theta: \theta \in \Theta\})$ are statistically equivalent, if there are two randomizations which transform P_θ into Q_θ for every $\theta \in \Theta$ and vice versa. It can be shown that this is the case, iff the distributions of the likelihood processes $(dP_\sigma/dP_\theta)_{\sigma \in \Theta}$ and $(dQ_\sigma/dQ_\theta)_{\sigma \in \Theta}$ w.r.t. P_θ and Q_θ, respectively, coincide for every $\theta \in \Theta$. Sometimes it is sufficient to consider only one fixed base point θ. For simplicity, we will do so in the following.

After these preliminaries, let us explain the basic idea of localization at hand of a simple example. We assume that E is the shift experiment generated by the standard Gaussian density φ, i.e.

$$\varphi(x) := \frac{1}{\sqrt{2\pi}} \exp(-x^2/2), \quad x \in \mathbb{R}^1 .$$

Choosing the base point $\theta = 0$ the corresponding n-fold product experiment $E^n = (\mathbb{R}^n, \mathcal{B}^n, \{P_\theta^n : \theta \in \mathbb{R}\})$ has the likelihood process

$$\frac{dP_\sigma^n}{dP_0^n}(x) = \exp(\sigma \sum_{i=1}^{n} x_i - \frac{n}{2}\sigma^2), \quad x \in \mathbb{R}^n, \quad \sigma \in \mathbb{R}^1, \quad (n \in \mathbb{N}) .$$

For different sample sizes $n \in \mathbb{N}$ the experiments E^n are not equivalent. Formally, this can be seen from the different distributions of the likelihood processes. However, it is also intuitively clear that increasing the sample size $n \in \mathbb{N}$ results in an increased information about the unknown parameter σ. If $n \to \infty$, this increase of information leads to the fact that for the limit experiment $E^{\mathbb{N}}$ the sequence of observations $\omega \in \Omega^{\mathbb{N}}$ completely determines the parameter σ. Now, we are going to carry out a rescaling of the parameters and put $\sigma := t/\sqrt{n}$. This gives a new sequence of experiments

$$E_n := (\mathbb{R}^n, \mathcal{B}^n, \{P_{t/\sqrt{n}}^n : t \in \mathbb{R}^1\}), \quad n \in \mathbb{N} .$$

The rescaling is called a <u>localization around the base point $\theta = 0$</u>. The appertaining likelihood processes are

$$\frac{dP^n_{t/\sqrt{n}}}{dP^n_o}(x) = \exp\left(\frac{t}{\sqrt{n}}\sum_{i=1}^{n} x_i - \frac{t^2}{2}\right), \quad x \in \mathbb{R}^n, \quad t \in \mathbb{R}^1, \quad (n \in \mathbb{N}).$$

Evidently, the distributions of the likelihood processes coincide showing that the experiments E_n are pairwise statistically equivalent.

Now, what is the practical significance of this rescaling? One task of the statistician is to choose a sample size large enough to admit sufficiently precise assertions about the parameter under consideration, and not too large to avoid unnecessary costs. Thereby, the problem arises of how to measure the dependance of precision and sample size. One possibility is a rescaling as described above. Since the rescaled experiments E_n, $n \in \mathbb{N}$, are pairwise equivalent, one can say that the influence of increasing the sample size is exactly compensated by the parameter-transformation. In other words, the rate of rescaling is a measure of the increase of precision which is achieved by enlarging the sample size. Let us be more explicit. Because of the equivalence of the E_n, $n \in \mathbb{N}$, a determination of the parameter t up to, say, two decimals is for all of these experiments subject to the same amount of stochastic uncertainty. For $n = 100$ this means a determination of the original parameter σ up to three decimals, for $n = 10\,000$ up to four decimals. The statistical precision increases like $\sqrt{n}$. In this sense the scale factor is, for fixed stochastic uncertainty, a measure for the statistical precision.

Now, we shall carry over the rescaling procedure to experiments which are more complicated than the Gaussian shift considered above. Let us first restrict ourselves to the simple case of a shift experiment generated by a sufficiently smooth and strictly positive univariate density h. The map $\ell(x) := -\log h(x)$, $x \in \mathbb{R}^1$, is called the likelihood contrast function. The likelihood processes belonging to the product experiments $(\mathbb{R}^n, \mathcal{B}^n, \{P^n_\theta : \theta \in \mathbb{R}^1\})$ are

$$\frac{dP^n_\sigma}{dP^n_o}(x) = \exp\left(-\sum_{i=1}^{n} (\ell(x_i-\sigma)-\ell(x_i))\right), \quad x \in \mathbb{R}^n, \quad \sigma \in \mathbb{R}^1, \quad (n \in \mathbb{N}).$$

Rescaling by $\sigma = t/\sqrt{n}$ as above and making a Taylor expansion of the resulting exponent yields

$$\frac{dP^n_{t/\sqrt{n}}}{dP^n_o} = \exp\left(\frac{t}{\sqrt{n}}\sum_{i=1}^{n} \ell'(x_i) - \frac{t^2}{2}\cdot\frac{1}{n}\sum_{i=1}^{n} \ell''(x_i) + R_n(x,t)\right),$$

$x \in \mathbb{R}^n$, $t \in \mathbb{R}^1$, $(n \in \mathbb{N})$. Under certain regularity conditions the remainders R_n vanish as n tends to infinity. Moreover, some other facts can be seen. By the law of large numbers, the random variables

$$x \longmapsto \frac{1}{n} \sum_{i=1}^{n} \ell''(x_i), \quad x \in \mathbb{R}^n ,$$

converge to a constant I , the Fisher's information. By the central limit theorem the random variables

$$x \longmapsto \frac{1}{\sqrt{n}} \sum_{i=1}^{n} \ell'(x_i), \quad x \in \mathbb{R}^n ,$$

are asymptotically normal with mean 0 and variance I . This Taylor expansion is the technical essence of the works of Wald, 1943, and LeCam, 1953. Under well-known regularity conditions it yields all the prominent limit theorems of asymptotic statistics. Refinements of such Taylor expansions play an important rôle within present research (cf. Pfanzagl, 1980). In contrast to the case of a Gaussian shift the rescaling does not necessarily result in pairwise equivalent experiments $E_n = (\mathbb{R}^n, \mathcal{B}^n, \{P^n_{t/\sqrt{n}}: t \in \mathbb{R}^1\})$, $n \in \mathbb{N}$. The distributions of the likelihood ratios do not coincide exactly. However, convergence of the exponents leads to a stabilization of the distributions. Roughly speaking, the likelihood processes are approximately of the form

$$\frac{dP^n_{t/\sqrt{n}}}{dP_o^n} \approx \exp(tX - \frac{t^2}{2} \cdot I), \quad t \in \mathbb{R}^1 , \ (n \in \mathbb{N}) ,$$

where I is a constant and X a centered Gaussian random variable with variance I . Note that the expression on the right is nothing else than the likelihood process of a Gaussian shift experiment. Hence, we may say that the likelihood processes of the localized experiments E_n , $n \in \mathbb{N}$, approximately coincide with the likelihood process of a Gaussian shift. In the light of the above-mentioned criterion for equivalence one is tempted to conclude that the locally reparametrized experiments E_n , $n \in \mathbb{N}$, are "approximately equivalent" to a Gaussian shift experiment.

The question occurs whether it is possible to give a mathematically precise meaning to the qualitative idea of "approximate equivalence". An affirmative answer to this question was given by LeCam, 1964. On the set of all experiment-types for a fixed parameter space he introduced a uniformizable topology $\mathcal{T}$ such that the available statistical precision is a continuous function: If two experiments E and F are close to each other in this topology, then the properties of a statistical procedure, say e.g. of corresponding Neyman-Pearson tests, do not differ very much under E and F . The concept of convergence w.r.t. this topology can be put into terms of likelihood processes. A sequence of experiments $(E_n)_{n \in \mathbb{N}}$ converges to an experiment E , iff the finite-dimensional marginal distributions of all likelihood processes of E_n converge weakly to the corresponding finite-dimensional marginal distributions of the likelihood processes of E .

As these remarks suggest, statistical problems on sequences of experiments may often be analyzed by increasing the sample size, thus "passing to the limit first and then arguing the case for the limiting problem" (LeCam, 1972). The present volume is concerned with the first step of this procedure. To illustrate this step, let us have a look at the case of shift experiments with smooth densities in the light of the above concept of convergence. From the Taylor expansion we see that the likelihood processes of the experiments E_n , $n \in \mathbb{N}$, converge in this sense to the likelihood process of a Gaussian shift,

$$\frac{dP^n_{t/\sqrt{n}}}{dP_o^n} \longrightarrow \exp(tX - \frac{t^2}{2} I) ,$$

meaning that the experiments E_n , $n \in \mathbb{N}$, can be approximated by this shift experiment in the topology $\mathcal{T}$. Hence, at least approximately, we have the same situation as in the case of a Gaussian shift considered above.

Summing up, the reparametrization results in a stabilization of the sequence of experiments insofar, as convergence to a Gaussian shift experiment was obtained. Now, it is sufficient to perform a statistical analysis of the limiting experiment. Then the results of asymptotic decision theory lead directly to the classical assertions of asymptotic statistics, e.g. to bounds on the efficiency of sequences of estimates. We do not dwell on this in any further detail, but refer to Lecam, 1972 and 1979.

It is an important problem to give conditions for the limit experiment to be a Gaussian shift and to judge whether this is somehow typical. This question threads through a considerable part of the work of LeCam. Already at an early stage it turned out that convergence to a Gaussian shift is closely related to differentiability properties of the original experiment (cf. LeCam, 1960). Another hint pointing at the fundamental rôle of Gaussian shift experiments can be obtained from the order of rescaling. In the presence of certain natural invariance conditions the order $n^{-1/2}$, $n \in \mathbb{N}$, is the slowest possible rate of reparametrization, and with this rescaling sequence the only possible limit experiments are Gaussian shifts (Strasser, 1984 a) .

Despite the predominance of approximations by Gaussian shifts it is not a priori clear that these approximations are always the most useful ones for practical purposes. As can be seen from the examples below there are quite a few cases where the same experiment can be embedded into different sequences of experiments with different limits. In those cases the usual argument in favour of a Gaussian shift approximation is that the statistical analysis of these limits is very easy; thus, the asymptotic assertions take a particular simple form in this case. But as long as there are no results as to which approximation is better, the usual approximation by Gaussian shift experiments has to be questioned. Let us give a few examples where different local approximations come into question.

<u>Example 1.</u> For every $p \in [0,1]$ let $B(p) := (1-p)\varepsilon_0 + p\varepsilon_1$, a two-point mass on $(\mathbb{R}^1, \mathcal{B}^1)$. Fix $p_0 \in (0,1)$ and consider the problem whether the true distribution $B(p)$ coincides with $B(p_0)$. Then the original experiment is $(\mathbb{R}^1, \mathcal{B}^1, \{B(p): p \in [0,1]\})$. A conventional Gaussian shift approximation is possible for $(\mathbb{R}^n, \mathcal{B}^n, \{B(p_0 + t/\sqrt{n})^n: t \in T_n\})$, where $T_n := \{t \in \mathbb{R}^1: p_0 + t/\sqrt{n} \in [0,1]\}$, $n \in \mathbb{N}$. This sequence of experiments converges to a Gaussian shift experiment and the approximation leads to the usual statistical procedures, which can also be obtained from the central limit theorem of DeMoivre and Laplace. But it is well-known that this method yields reasonable results only in case p_0 lies close to the center of $[0,1]$. If p_0 is close to 0 or 1, then another approximation gives much better results. For small p_0 let $F_n := (\mathbb{R}^n, \mathcal{B}^n, \{B(\lambda/n)^n: \lambda \in T_n\})$, where $T_n := \{\lambda \geq 0: \lambda/n \leq 1\}$, $n \in \mathbb{N}$. This sequence converges to a Poisson experiment. Fixing a sample size $n \in \mathbb{N}$ and putting $\lambda_0 := np_0$, the testing of p_0 amounts to the testing of λ_0 for the experiment F_n. It is still an open question for which combinations of n and p_0 which of both approximations is preferable.

<u>Example 2.</u> Let e_i, $1 \leq i \leq m$, denote the natural ON-base of $\mathbb{R}^m$, and consider the open unit simplex

$$\overset{\circ}{S}_m := \{p \in (0,1)^m : \sum_{i=1}^{m} p_i = 1\} .$$

Let $M(p) := \sum_{i=1}^{n} p_i \varepsilon_{e_i}$, $p \in \overset{\circ}{S}_m$. Now, fix $p^{(o)} = (p_1^{(o)}, \ldots, p_m^{(o)})$ and consider the problem whether the true distribution $M(p)$ coincides with $M(p^{(o)})$. Then, the original experiment is $(\mathbb{R}^m, \mathcal{B}^m, \{M(p): p \in \overset{\circ}{S}_m\})$. Let $T := \{t \in \mathbb{R}^m : t_1 + \ldots + t_m = 0\}$. A conventional Gaussian shift approximation is possible for

$$(\mathbb{R}^{m \cdot n}, \mathcal{B}^{m \cdot n}, \{M(p^{(o)} + t/\sqrt{n})^n: t \in T_n\})$$

where $T_n := \{t \in T: p^{(o)} + t/\sqrt{n} \in \overset{\circ}{S}_m\}$, $n \in \mathbb{N}$. This sequence of experiments converges to a Gaussian shift experiment; hence, the approximation leads to the usual χ^2-procedures (for details see Strasser, 1984 b, Sec. 82). In the light of Example 1 it seems to be questionable whether this method also renders good results, if some of the $p_i^{(o)}$, $1 \le i \le m$, are close to zero or close to one. For such coordinates a reparametrization of the form $p_i = \lambda_i/n$ or $p_i = 1 - \lambda_i/n$ might be more appropriate. Then, the rescaling rates could be different for different coordinates. Such mixed reparametrizations would produce limit experiments which are products of a Gaussian shift and a Poisson experiment. To our knowledge, such approximations have not yet been investigated, although it is a well-known fact that the χ^2-method performs not too well, if some of the $p_i^{(o)}$ are small.

In the case of both just described examples the Gaussian shift approximation is inferior to some other approximation, if the parameter under consideration is close to the boundary of the parameter space. Here, the meaning of "close to" depends upon the sample size. For every parameter, however close it might be to the boundary, the Gaussian shift approximation - if possible at all - is superior to any other approximation, provided that the sample size is sufficiently large. However, for small sample sizes it often seems more reasonable to embed the experiment into a sequence for which the underlying probability distributions tend to the boundary of the experiment. These heuristic aspects are intuitively clear, and can be supported by numerical studies for binary experiments.

The following example is fundamental in nature.

<u>Example 3.</u> Let $(\mathbb{R}^1, \mathcal{B}^1, \{P_\theta: \theta \in \mathbb{R}^1\})$ be a shift experiment generated by a univariate Lebesgue density h. Before entering the case of smooth densities we consider the non-regular case of the uniform density $h := \frac{1}{2} \cdot 1_{(-1,1)}$. Let $\theta_o \in \mathbb{R}^1$ be the value of statistical interest. Then, the sequence $(\mathbb{R}^n, \mathcal{B}^n, \{P^n_{\theta_o + t/\sqrt{n}}: t \in \mathbb{R}^1\})$, $n \in \mathbb{N}$, converges to the experiment

$$F := (\mathbb{R}^2, \mathcal{B}^2, \{Q_t: t \in \mathbb{R}^1\}),$$

defined by

$$\frac{dQ_t}{d\lambda^2}(x,y) := e^{x-y} \cdot 1_{(-\infty,t) \times (t,\infty)}(x,y), \quad (x,y) \in \mathbb{R}^2, \quad t \in \mathbb{R}^1 .$$

Using the terminology of this book, F is a Poisson experiment. In this case an approximation by a Gaussian shift experiment is impossible.

Now, we want to show that there are examples with smooth densities h, for which - besides the Gaussian shift approximation - there are also approximations by Poisson experiments like F. Our construction will be a little bit artificial. Nevertheless, it illustrates the possibility of such situations.

Example 4. For $\sigma > 0$ let

$$h_\sigma(x) := C(\sigma)(1 - \Phi(\frac{x^2-1}{\sigma})), \ x \in \mathbb{R}^1,$$

and $P_{\sigma,\theta}$ the probability measure on $(\mathbb{R}^1, \mathcal{B}^1)$ with Lebesgue density $x \longrightarrow h_\sigma(x-\theta)$, $x \in \mathbb{R}^1$, $(\theta \in \mathbb{R}^1)$. ($\Phi$ denotes the standard normal distribution function.) We observe that

$$\lim_{\sigma \to 0} h_\sigma(x) = \frac{1}{2} \cdot 1_{(-1,1)}(x) =: h(x), \ x \in \mathbb{R}^1.$$

Figure 1

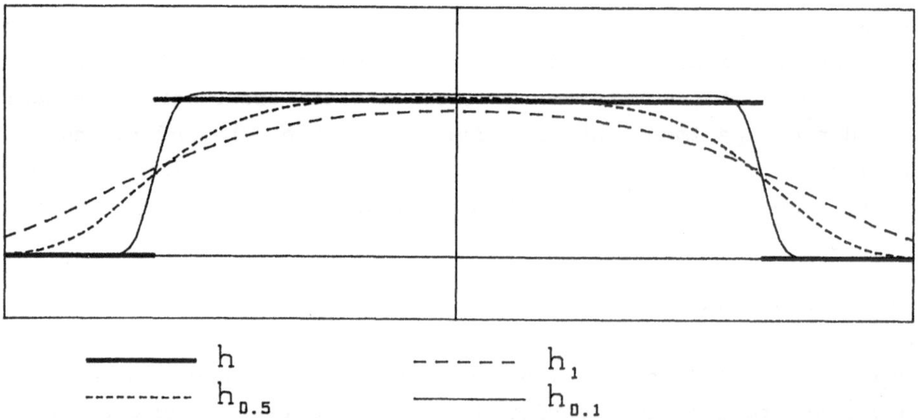

On the other hand, for every $\sigma > 0$ the density h_σ is smooth and the experiment $(\mathbb{R}^1, \mathcal{B}^1, \{P_{\sigma,\theta}: \theta \in \mathbb{R}^1\})$ admits an approximation by a Gaussian shift experiment. To be more precise, the sequence

$$(\mathbb{R}^n, \mathcal{B}^n, \{P_{\sigma,\theta_0 + t/\sqrt{n}}: \theta \in \mathbb{R}^1\}), \ n \in \mathbb{N},$$

converges to a Gaussian shift for every $\sigma > 0$. However there is still another conceivable approximation: The experiments

$$E_n := (\mathbb{R}^n, \mathcal{B}^n, \{P_{1/n,\ \theta_0 + t/n}: t \in \mathbb{R}\}), \ n \in \mathbb{N},$$

tend to the Poisson experiment F. There are good arguments that for small σ the embedding of our problem into the sequence (E_n) gives the better approximation.

As to the investigation of other than Gaussian shift approximations sometimes the objection is raised that for applications all problems could be modelled by sufficiently regular densities. Such regular models would then be approximable by Gaussian shift experiments. But in view of Example 4 just the opposite standpoint seems to be possible. Even in the case of regular distributions it might be preferable to use other than Gaussian shift approximations, namely if the sample size is small and the distributions are close to irregular ones. E.g. if the densities have large derivatives at some points, then one might prefer to embed the experiment into a sequence the densities of which finally have jumps at these points (see Figure 1). A similar approach is conceivable if the moments of the distributions are large.

At this point, we have to make mention of the fact that the insight in certain disadvantages of the Gaussian shift approximation is not new at all. Within recent years it led to a highly developed theory which is based on an approach totally different from the one we propose. We refer to the theory of asymptotic expansions which has, above all, been developed by Pfanzagl and his school. Its basic idea is to refine the Taylor expansion of the exponent of the likelihood process which we have mentioned above. Instead of approximating the localized sequence of experiments

$$E_n = (\Omega^n, A^n, \{P^n_{\theta + t/\sqrt{n}}: t \in \sqrt{n}\,(\Theta-\theta)\})\ ,\ n \in \mathbb{N}\ ,$$

by the limiting Gaussian shift experiment, one then obtains an accompanying sequence of curved exponential experiments which approximates $(E_n)_{n \in \mathbb{N}}$ of considerably higher order. To our mind, it is a challenging question whether near the boundary of regular experiments the method of asymptotic expansions or our proposal leads to better approximations.

The above remarks were intended to show the importance of taking into consideration also other than classical approximations by Gaussian shift experiments. We hope to stimulate the reader's interest for problems like this. With the present textbook we want to provide the necessary mathematical tools and a readable basis for further research. However, we do not intend to touch upon the above-raised questions as to the numerical performance of the alternative Poisson approximations. These problems are open and could be subject to future research.

Now, we briefly summarize the contents of the present volume. Its first part surveys and supplements LeCam's theory of infinitely divisible experiments, thus incorporating new facts as well as detailed proofs of some known results. It starts from the representation theory for Gaussian and Poisson experiments, the latter being introduced via compound Poisson experiments. In particular, the interplay between the concepts of Lévy measures on simplices and of Poisson experiments is discussed thoroughly. The results obtained there are then applied to show that any weak accumulation point of an infinitesimal and bounded triangular array of experiments has a unique representation as a direct product of a Gaussian and a Poisson factor. This leads to a "Lévy-Khintchine representation" of the Hellinger transforms of infinitely divisible experiments. Finally, necessary and sufficient conditions are given, under which all accumulation points of an infinitesimal and bounded triangular array are either pure Gaussian or pure Poisson experiments, respectively have a compound Poisson factor.

In the second chapter we present a different approach to infinitely divisible experiments, based on semi-groups and the theory of infinitely divisible distributions. We first deal with the case of a finite parameter set Θ. Then an experiment $E = (\Omega, A, \{P_\theta : \theta \in \Theta\})$ is infinitely divisible iff the distributions of all log-likelihood processes $(\log \frac{dP_\sigma}{dP_\theta})_{\sigma \in \Theta \setminus \{\theta\}}$ w.r.t. P_θ are infinitely divisible on $([-\infty,+\infty)^{\Theta \setminus \{\theta\}},+)$, $\theta \in \Theta$. Passing to a standard representation, infinitely divisible experiments can be described by families of infinitely divisible distributions on the unit simplex $S_\Theta := \{z \in [0,1]^\Theta : \sum_{\theta \in \Theta} z_\theta = 1\}$. Let $(E_t)_{t \geq 0}$ denote the continuous semi-group induced by E. This semi-group admits a Lévy-Khintchine representation, given by a Gaussian part and a Lévy measure. The Gaussian part can be obtained from the Gaussian parts of the log-likelihood processes. The Lévy measure has an interpretation in terms of the semi-group induced by E: It can be viewed as the "derivative" of the standard measures of E_t, $t \geq 0$, at the origin w.r.t. the vague topology on $S_\Theta \setminus \{e_\Theta\}$, $e_\Theta := (\frac{1}{|\Theta|}, \ldots, \frac{1}{|\Theta|})$. The semi-group (E_t) is uniquely determined by its generating functional which is an almost positive operator.

Next, we consider arbitrary parameter sets. Then infinitely divisible experiments can be described by a projective limit of continuous convolution semi-groups defined on a projective system of simplices. A Lévy-Khintchine formula can be derived for each semi-group. Again, the Lévy-measures are the derivatives of the semigroups w.r.t. vague convergence.

In the third part we show that our concept of Poisson experiments coincides with the one of LeCam, 1974. According to LeCam a standard Poisson experiment consists of distributions of generalized Poisson processes given a parametrized family of intensities. Relying on a Consistency theorem of Bochner, 1955, the existence of a generalized Poisson process given an arbitrary content as intensity is proved. Employing this result and a fundamental lemma of Janssen, it is then shown that every Poisson experiment in our sense is equivalent to a standard Poisson experiment in the sense of LeCam. Conversely, every standard Poisson experiment is a Poisson experiment; this is proved relying on the convergence theory for Lévy measures developed in Part I .

The starting point of the fourth part are those non-regular situations in asymptotic statistics, where the densities may have jumps. Such situations have been considered under a couple of conditions by Ibragimov and Has'minskii, 1972 and 1981, and by Pflug, 1983. We treat the problem by LeCam's theory of infinitely divisible experiments. It is shown that the limit experiments obtained are particular cases of what we call experiments with independent increments. Necessary and sufficient conditions are obtained for the weak convergence of triangular arrays to experiments with independent increments. The class of all experiments with independent increments is described in terms of both, their Hellinger transforms and the associated semi-groups of binary experiments. In particular, it is shown that an experiment with independent increments is completely determined by its binary subexperiments. Finally, a simple set of conditions is given which implies the situation considered by Ibragimov and Has'minskii.

I. LIMITS OF TRIANGULAR ARRAYS OF EXPERIMENTS

Hartmut Milbrodt and Helmut Strasser

1. Basic Concepts

An experiment for a non-empty parameter set T is a triplet $E = (\Omega, \mathcal{A}, \{P_t: t \in T\})$ consisting of a measurable space $(\Omega, \mathcal{A})$ and a family of probability measures on $\mathcal{A}$. The notion of an experiment is as fundamental for the statistical theory as the notion of a stochastic process is for probability theory. For many purposes probability theory considers stochastic processes as equivalent if their distributions coincide. Moreover, the notion of weak convergence of stochastic processes only depends on the equivalence classes characterized by their distributions. A similar situation arises in the theory of experiments. For motivation we explain some basic ideas at hand of binary experiments. For the general case we refer to the literature.

Let $T := \{0,1\}$. An experiment for this parameter space is called a binary experiment. Binary experiments $E = (X, \mathcal{A}, \{P_0, P_1\})$ and $F = (Y, \mathcal{B}, \{Q_0, Q_1\})$ can be compared as follows: E is ε-better than F, $\varepsilon \geq 0$, if for every critical function ψ on $(Y, \mathcal{B})$ there exists a critical function φ on $(X, \mathcal{A})$ such that

$$P_0\varphi \leq Q_0\psi + \varepsilon \quad \text{and} \quad P_1\varphi \geq Q_1\psi - \varepsilon .$$

If $\Delta(E,F)$ denotes the infimum over all $\varepsilon \geq 0$ for which E is ε-better than F and vice versa, then Δ is a pseudo-distance on the class of all binary experiments. Now, it is obvious how to define equivalence and convergence of binary experiments. Easy applications of the Neyman-Pearson theory show that $\Delta(E,F) = 0$ iff the distributions $\mathcal{L}(\frac{dP_1}{dP_0} \mid P_0)$ and $\mathcal{L}(\frac{dQ_1}{dQ_0} \mid Q_0)$ coincide. Similarly, if $E_n = (X_n, \mathcal{A}_n, \{P_{no}, P_{n1}\})$, $n \in \mathbb{N}$, is a sequence of binary experiments then $\Delta(E_n, E) \longrightarrow 0$ iff

$$\mathcal{L}(\frac{dP_{n1}}{dP_{no}} \mid P_{no}) \longrightarrow \mathcal{L}(\frac{dP_1}{dP_0} \mid P_0), \quad \text{weakly} .$$

The point is that these notions can be generalized to experiments with arbitrary parameter sets. The decision theoretic formulation of equivalence and weak convergence of experiments was developed by LeCam, 1964 and 1972. From LeCam, 1974, it is known that equivalence and weak convergence can be phrased in terms of the likelihood processes of the experiments, as we do below. A proof of these facts can be found e.g. in Strasser, 1984 b.

Suppose that $E = (\Omega, \mathcal{A}, \{P_t: t \in T\})$ is an experiment. The likelihood process of E with base $s \in T$ is $\left(\frac{dP_t}{dP_s}\right)_{t \in T}$, the distribution taken with respect to P_s.

(1.1) Definition. Two experiments $E = (\Omega_1, \mathcal{A}_1, \{P_t: t \in T\})$ and $F = (\Omega_2, \mathcal{A}_2, \{Q_t: t \in T\})$ are equivalent $(E \sim F)$ if

$$\mathcal{L}\left(\left(\frac{dP_t}{dP_s}\right)_{t \in T} \Big| P_s \right) = \mathcal{L}\left(\left(\frac{dQ_t}{dQ_s}\right)_{t \in T} \Big| Q_s \right), \quad s \in T .$$

The respective equivalence classes are called experiment-types. Convergence and limits of experiments will always be understood in the sense of the following definition.

(1.2) Definition. A net of experiments $E_\nu = (\Omega_\nu, \mathcal{A}_\nu, \{P_{\nu t}: t \in T\})$, $\nu \in N$, converges weakly to $E = (\Omega, \mathcal{A}, \{P_t: t \in T\})$ if for every finite subset α of T and every $s \in \alpha$

$$\mathcal{L}\left(\left(\frac{dP_{\nu t}}{dP_{\nu s}}\right)_{t \in \alpha} \Big| P_{\nu s} \right) \longrightarrow \mathcal{L}\left(\left(\frac{dP_t}{dP_s}\right)_{t \in \alpha} \Big| P_s \right), \text{ weakly.}$$

Note that this concept of convergence does only depend on the experiment-types under consideration. We refrain from an explicit description of the topology belonging to the notion of weak convergence. Topological concepts will be interpreted in terms of nets. Thus, the well-known fact that the set of experiment-types for any fixed parameter space is compact (LeCam, 1974) will only be employed for extracting convergent subnets from nets of experiments.

The analytical tool for handling experiments is the Hellinger transform.

Let $A(T)$ be the class of all finite non-empty subsets of T. If $E = (\Omega, \mathcal{A}, \{P_t: t \in T\})$ and $\alpha \in A(T)$ let $E_\alpha := (\Omega, \mathcal{A}, \{P_t: t \in \alpha\})$ be the <u>restriction</u> of E to α . For every $\alpha \in A(T)$ denote $S_\alpha :=$ $\{(x_t)_{t\in\alpha} \in [0,1]^\alpha : \sum_{t\in\alpha} x_t = 1\}$, and $\overset{\circ}{S}_\alpha := S_\alpha \cap (0,1)^\alpha$. If $\nu|A$ is a σ-finite measure dominating $\{P_t : t\in\alpha\}$ then

$$H(E_\alpha)(z) := \int \prod_{t\in\alpha} \left(\frac{dP_t}{d\nu}\right)^{z_t} d\nu \ , \ z \in S_\alpha \ ,$$

is independent of ν , and $z \longmapsto H(E_\alpha)(z)$ is called the <u>Hellinger transform</u> of E_α (here $0^0 = 1$). For every pair $(s,t) \in T^2$ the <u>Hellinger distance</u> of P_s and P_t is given by

$$d(P_s,P_t) := \left(1 - H(E_{\{s,t\}})\left(\tfrac{1}{2},\tfrac{1}{2}\right)\right)^{1/2} .$$

Note that $H(E_\alpha)$ is continuous on $\overset{\circ}{S}_\alpha$, $\alpha \in A(T)$; if E is <u>homogeneous</u> , i.e. if the probability measures defining E are pairwise equivalent, then $H(E_\alpha)$ is continuous on the whole of S_α .

Let E and F be experiments for the parameter set T . It is known from LeCam, 1972, that $E_\alpha \sim F_\alpha$ iff $H(E_\alpha) = H(F_\alpha)$, $\alpha \in A(T)$. Since $E \sim F$ iff $E_\alpha \sim F_\alpha$ for every $\alpha \in A(T)$, it follows that the equivalence class of E is completely determined by the system $H(E_\alpha)$, $\alpha \in A(T)$. The systems $(E_\alpha)_{\alpha\in A(T)}$ and $(H(E_\alpha))_{\alpha\in A(T)}$ are both projective: $\alpha \leq \beta$ implies $(E_\beta)_\alpha \sim E_\alpha$ and $H(E_\beta)|_{S_\alpha} = H(E_\alpha)$. Conversely, if $(E_{(\alpha)})_{\alpha\in A(T)}$ is a projective system of experiments, then there exists a projective limit, i.e. an experiment E for the parameter space T satisfying $E_{(\alpha)} \sim E_\alpha$, $\alpha \in A(T)$. This was noted by LeCam, 1972.

Let E_n , $n \in \mathbb{N}$, be experiments for the parameter space T . Then it is also known from LeCam, 1972, that $E_{n,\alpha} \longrightarrow E_\alpha$ iff $H(E_{n,\alpha}) \longrightarrow$ $H(E_\alpha)$, pointwise on S_α , $\alpha \in A(T)$. It follows that $E_n \longrightarrow E$ weakly iff $H(E_{n,\alpha}) \longrightarrow H(E_\alpha)$ for every $\alpha \in A(T)$. Moreover, due to the compactness-assertion cited above, $(E_n)_{n\in\mathbb{N}}$ converges weakly iff $(H(E_{n,\alpha}))_{n\in\mathbb{N}}$ converges for every $\alpha \in A(T)$.

Experiments $E = (\Omega_1, A_1, \{P_t : t\in T\})$ and $F = (\Omega_2, A_2, \{Q_t : t\in T\})$ can be multiplied as usual to obtain their <u>direct product</u>

$$E \otimes F := (\Omega_1 \times \Omega_2, A_1 \otimes A_2, \{P_t \otimes Q_t : t\in T\}) .$$

Apparently, $H((E\otimes F)_\alpha) = H(E_\alpha)\cdot H(F_\alpha)$, $\alpha \in A(T)$. If $E_n =$ $(\Omega_n, A_n, \{P_{nt} : t\in T\})$, $n \in \mathbb{N}$, are experiments and $(a_n)_{n\in\mathbb{N}} \subset [0,1]$ is such that $\sum_{n=1}^{\infty} a_n = 1$, then the <u>direct convex combination</u>

$$\bigoplus_{n=1}^{\infty} a_n E_n := \left(\bigoplus_{n=1}^{\infty} \Omega_n, \bigoplus_{n=1}^{\infty} A_n, \left\{\bigoplus_{n=1}^{\infty} a_n P_{nt} : t\in T\right\}\right)$$

is explained as follows. As sample space $\bigoplus_{n=1}^{\infty} \Omega_n$ take the free union of the Ω_n , $n \in \mathbb{N}$; the σ-field $\bigoplus_{n=1}^{\infty} \mathcal{A}_n$ and the probability measures $\bigoplus_{n=1}^{\infty} a_n P_{nt}$, $t \in T$, are defined by the requirements that their traces on each Ω_n equal $\mathcal{A}_n$ and $a_n \cdot P_{nt}$, $t \in T$, respectively. Note that

$$H((\bigoplus_{n=1}^{\infty} a_n E_n)) = \sum_{n=1}^{\infty} a_n H(E_{n,\alpha}), \; \alpha \in A(T).$$

In the present chapter we deal with weak convergence of product experiments. Let $(k_n)_{n \in \mathbb{N}} \subset \mathbb{N}$ be a sequence satisfying $k_n \uparrow \infty$. For every $n \in \mathbb{N}$ consider experiments

$$E_{ni} = (\Omega_{ni}, \mathcal{A}_{ni}, \{P_{nit}: t \in T\}), \quad 1 \leq i \leq k_n .$$

The double sequence $(E_{ni})_{1 \leq i \leq k_n, \, n \in \mathbb{N}}$ is called a <u>triangular array of experiments</u>. For convenience we denote

$$E_n := (\prod_{i=1}^{k_n} \Omega_{ni}, \bigotimes_{i=1}^{k_n} \mathcal{A}_{ni}, \{\bigotimes_{i=1}^{k_n} P_{nit}: t \in T\}), \quad n \in \mathbb{N} .$$

The most important special case is that of independent and identically distributed observations. In order to cover also the non-identically distributed case, the present general set-up was chosen. It occurs e.g. in case of sequences of regression problems. To be specific, let Q be a probability measure on $(\mathbb{R}^1, \mathcal{B}(\mathbb{R}^1))$, $k \in \mathbb{N}$, and $C_n = (c_{n1}, \ldots, c_{nk_n})$ be $k_n \times k$ design matrices, $n \in \mathbb{N}$. Let Q_s denote the translate of Q under $s \in \mathbb{R}^1$. Then $(\mathbb{R}^1, \mathcal{B}(\mathbb{R}^1), \{P_{nit}: t \in \mathbb{R}^k\})$ where $P_{nit} := Q_{\langle c_{ni}, t \rangle}$, $1 \leq i \leq k_n$, $n \in \mathbb{N}$ is a triangular array of experiments. Examples of such arrays will be considered in Sections 5 and 6.

Now, we return to the more general situation. The problem is to give conditions for the weak convergence of the sequence $(E_n)_{n \in \mathbb{N}}$, and to identify the limit. For binary experiments this problem was treated in full generality for the first time by LeCam, 1969. The case of an arbitrary parameter set was considered in LeCam, 1974. An elaboration and supplementation of this case is given here. We do not aim at a complete documentation of priorities since every essential idea goes back to LeCam.

Different to LeCam we shall impose a certain boundedness condition on the triangular array. This guarantees that each of the accumulation points of $(E_n)_{n \in \mathbb{N}}$ is <u>pairwise imperfect</u> in the sense that no two of its underlying probability measures are disjoint. The restriction is not severe, since the general case may be reduced to this one by a suitable decomposition of the parameter set (LeCam, 1974, and Part II, Sec. 9 of this volume). On the other hand it ensures that the problem can be solved within the framework of the theory of Lévy measures on simplices which is developed here. This turns out to be especially useful in Sec. 6 where criteria for normal and (compound) Poisson convergence are first obtained in terms of Lévy measures and then rephrased in terms of the likelihood ratios appertaining to E_{ni}, $1 \leq i \leq k_n$, $n \in \mathbb{N}$.

2. Gaussian Experiments

If H is a finite-dimensional Hilbert space, there exists a standard Gaussian measure N_H on the Borel-σ-field $\mathcal{B}(H)$ of H; let ε_x denote the Dirac-measure sitting at $x \in H$.

(2.1) Definition. A Gaussian shift experiment on a finite-dimensional Hilbert space H is an experiment for the parameter space H which is equivalent to the experiment $(H, \mathcal{B}(H), \{N_H * \varepsilon_x : x \in H\})$.

Elementary calculations using quadratic completion show that the Hellinger transforms of a finite-dimensional Gaussian shift E are given by

$$H(E_\alpha)(z) = \exp\{\tfrac{1}{2}\,(\sum_{s,t\in\alpha} z_s z_t \langle s,t\rangle - \sum_{t\in\alpha} z_t \|t\|^2)\},\quad \alpha \in A(H),\ z \in S_\alpha.$$

Gaussian shift experiments on finite-dimensional Hilbert spaces are the most important examples of limit experiments arising in the classical theory of asymptotic statistics. We do not dwell on this any further, but turn to the infinite-dimensional case.

(2.2) Definition. A Gaussian shift experiment on an arbitrary Hilbert space H is an experiment for the parameter space H whose restrictions to the finite-dimensional subspaces of H are Gaussian shifts.

Gaussian shift experiments on infinite dimensional Hilbert spaces play a rôle in asymptotic statistics of non-parametric problems. Compare e.g. Moussatat, 1976, and Millar, 1979.

For our purposes we need the concept of general Gaussian experiments. Let $T \neq \emptyset$ be an arbitrary set.

(2.3) Definition. Let $E = (\Omega, A, \{P_t : t \in T\})$ be an experiment. Then E is a Gaussian experiment if it is a subexperiment of a Gaussian shift. To be precise: If there exists a Hilbert space $(H, \langle\cdot,\cdot\rangle)$ and a map $\psi: T \longrightarrow H$ such that

$$E \sim (\Omega_1, A_1, \{Q_{\psi(t)} : t \in T\}),$$

where $(\Omega_1, A_1, \{Q_x : x \in H\})$ is any Gaussian shift on H.

The preceding definition serves mainly as a motivation of the concept. It becomes analytically tractable by the following lemma.

(2.4) Lemma. An experiment E for the parameter space T is a Gaussian experiment iff there exists a positive semi-definite and symmetric kernel $K: T^2 \longrightarrow \mathbb{R}^1$ such that

$$H(E_\alpha)(z) = \exp\left[\frac{1}{2}\left(\sum_{s,t\in\alpha} z_s z_t K(s,t) - \sum_{t\in\alpha} z_t K(t,t)\right)\right],$$

whenever $\alpha \in A(T)$, $z \in S_\alpha$.

Proof. (1) Assume that E is a Gaussian experiment. Let $F = (\Omega_1, A_1, \{Q_x: x \in H\})$ be a Gaussian shift on some Hilbert space $(H, <\cdot,\cdot>)$ such that

$$E \sim (\Omega_1, A_1, \{Q_{\psi(t)}: t \in T\}).$$

Taking $K(s,t) = <\psi(s), \psi(t)>$, $s \in T$, $t \in T$, yields the desired representation of the Hellinger transforms.

(2) Conversely, assume that the Hellinger transforms have the above-stated form with some kernel K. It is well-known that for every positive semi-definite and symmetric kernel $K: T \times T \longrightarrow \mathbb{R}^1$ there exists a Hilbert space $(H, <\cdot,\cdot>)$ and a map $\psi: T \longrightarrow H$ such that $K(s,t) = <\psi(s), \psi(t)>$, $s \in T$, $t \in T$. Then, for every $\beta \in A(H)$ we have

$$H(F_\beta)(z) = \exp\left[\frac{1}{2}\left(\sum_{x,y\in\beta} z_x z_y <x,y> - \sum_{x\in\beta} z_x <x,x>\right)\right],$$

whenever $z \in S_\beta$. It follows that the Hellinger transforms of E and of $(\Omega_1, A_1, \{Q_{\psi(t)}: t \in T\})$ coincide. □

(2.5) Examples.

(1) A Gaussian shift experiment on a Hilbert space $(H, <\cdot,\cdot>)$ is a Gaussian experiment with kernel $<\cdot,\cdot>$. All Gaussian shift experiments with $H = \mathbb{R}^1$ are given by the kernels $K: (s,t) \longmapsto a\,s\,t$ $(a > 0)$.

(2) Let $T = \mathbb{R}^1$ and $0 < \rho \leq 2$. Then

$$K: (s,t) \longmapsto a\,(|s|^\rho + |t|^\rho - |s-t|^\rho) \quad (a > 0)$$

is a positive definite and symmetric kernel. If $\rho = 2$, then this reduces to the situation of (1). If $\rho < 2$, the corresponding

Gaussian experiments arise e.g. as weak limits of experiments $(\mathbb{R}^n, \mathbb{B}^n, \{P_{n^{-1/\alpha}t} : t \in \mathbb{R}^1\})$, $n \in \mathbb{N}$, where

$$\frac{dP_\theta}{d\lambda^1}(x) := C(\rho) \cdot \exp(-|x-\theta|^\rho), \quad x \in \mathbb{R}^1, \quad \theta \in \mathbb{R}^1,$$

and $\alpha := \frac{\rho-1}{2}$. According to LeCam, 1969 (p. 109) this situation is " ... plus compliquée mais plus interessante ". It has been studied by several authors, e.g. Pflug, 1982, and Strasser, 1984 a. In the latter paper it has been shown that, under some additional invariance conditions, all kernels on $\mathbb{R}^1$ are of the form given above.

From the proof of the preceding lemma it is clear that for every positive semi-definite and symmetric kernel $K: T^2 \longrightarrow \mathbb{R}^1$ there exists a Gaussian experiment having Hellinger transforms of the form given in the lemma. Obviously, two Gaussian experiments, whose Hellinger transforms can be represented by the same kernel, are equivalent. The converse, however, is not valid.

<u>(2.6) Lemma.</u> Two positive semi-definite and symmetric kernels $K_i: T^2 \longrightarrow \mathbb{R}^1$, $i = 1,2$, define equivalent Gaussian experiments iff for every pair $(s,t) \in T^2$

$$K_1(s,t) - \frac{1}{2}(K_1(s,s) + K_1(t,t)) = K_2(s,t) - \frac{1}{2}(K_2(s,s) + K_2(t,t)).$$

<u>Proof.</u> The condition is necessary: If $s = t$ it is trivially satisfied; if $s \neq t$, it follows from the coincidence of the Hellinger transforms for $\alpha = \{s,t\}$ and $z = (\frac{1}{2}, \frac{1}{2})$. Conversely, let the condition be satisfied. Then for every $\alpha \in A(T)$ and $z \in S_\alpha$

$$\sum_{s,t \in \alpha} z_s z_t K_i(s,t) - \sum_{t \in \alpha} z_t K_i(t,t) =$$

$$= \sum_{s,t \in \alpha} z_s z_t K_i(s,t) - \frac{1}{2} \sum_{s \in \alpha} z_s K_i(s,s) - \frac{1}{2} \sum_{t \in \alpha} z_t K_i(t,t) =$$

$$= \sum_{s,t \in \alpha} z_s z_t \left(K_i(s,t) - \frac{1}{2}(K_i(s,s) + K_i(t,t))\right),$$

$i = 1,2$. Hence, the Hellinger transforms coincide. □

Let us call two kernels <u>equivalent</u> if they define equivalent Gaussian experiments.

In the preceding proof we observed that for a Gaussian experiment $(\Omega, A, \{P_t: t \in T\})$ with kernel K

$$1 - d^2(P_s, P_t) = H(E_{\{s,t\}})(\tfrac{1}{2}, \tfrac{1}{2}) = \exp\left[\tfrac{1}{4}\left(K(s,t) - \frac{K(s,s)+K(t,t)}{2}\right)\right]$$

for every pair $(s,t) \in T^2$. This yields the following important

<u>(2.7) Theorem.</u> Two Gaussian experiments $(\Omega_i, A_i, \{P_{it}: t \in T\})$, $i = 1,2$, are equivalent iff

$$d(P_{1s}, P_{1t}) = d(P_{2s}, P_{2t}) \quad \text{for all pairs } (s,t) \in T^2 .$$

In order to achieve uniqueness of the kernel defining a Gaussian experiment we use a suitable standardization.

<u>(2.8) Lemma and Definition.</u> Let $t_o \in T$. Then each equivalence class of kernels contains exactly one kernel K_{t_o} satisfying $K_{t_o}(t,t_o) = K_{t_o}(t_o,t) = 0$ for all $t \in T$. K_{t_o} is called the <u>kernel standardized at t_o</u>.

<u>Proof.</u> If K is an arbitrary kernel then

$$(s,t) \longmapsto K(s,t) - K(s,t_o) - K(t,t_o) + K(t_o,t_o), \quad (s,t) \in T^2$$

is an appropriately standardized equivalent kernel. Moreover, if K_1 and K_2 are equivalent kernels both being standardized at t_o, then

$$K_1(s,t_o) - \tfrac{1}{2}(K_1(s,s) + K_1(t_o,t_o)) =$$

$$= K_2(s,t_o) - \tfrac{1}{2}(K_2(s,s) + K_2(t_o,t_o)), \quad s \in T,$$

implies $K_1(s,s) = K_2(s,s)$ for all $s \in T$. Now, from the equivalence condition it follows that $K_1 = K_2$. □

<u>(2.9) Discussion.</u> Let $E = (\Omega, A, \{P_t: t \in T\})$ be a Gaussian experiment and $t_o \in T$. In view of the preceding lemma it makes sense to speak of the kernel of E standardized at t_o. Apparently, it can be obtained from any kernel defining the Hellinger transforms of E. Moreover, it can be obtained from the Hellinger distances as the following argument shows. Denote

$$a(s,t) := -\log(1 - d^2(P_s, P_t)), \quad (s,t) \in T^2 .$$

Then the kernel standardized at $t_o \in T$ is

$$\overline{K}(s,t) = 4(a(s,t_o) + a(t,t_o) - a(s,t)), \quad (s,t) \in T^2.$$

Indeed, if K is any kernel of E, then

$$a(s,t) = -K(s,t) + \frac{K(s,s)+K(t,t)}{2}, \quad (s,t) \in T^2,$$

which yields

$$\overline{K}(s,t) = K(s,t) - K(s,t_o) - K(t,t_o) + K(t_o,t_o) = K_{t_o}(s,t), \quad (s,t) \in T^2.$$

The standardized kernels are related to the likelihood processes of Gaussian experiments, as we shall see below.

<u>(2.10) Lemma.</u> Every Gaussian experiment is homogeneous.

<u>Proof.</u> Every Gaussian shift is homogeneous. □

Since for homogeneous experiments the likelihood processes take values in $(0,\infty)$, it is convenient to consider the log-likelihood processes of a Gaussian experiment. The following theorem shows that the notion of Gaussian experiments introduced in LeCam, 1974, coincides with the one employed here.

<u>(2.11) Theorem.</u> Let $E = (\Omega, A, \{P_t : t \in T\})$ be an experiment. The following assertions are equivalent:

(1) E is a Gaussian experiment.

(2) E is homogeneous, and every log-likelihood process of E is a Gaussian process.

(3) E is homogeneous and at least one log-likelihood process is Gaussian.

<u>Proof.</u> Let us first show that (1) implies (2). Let E be Gaussian, $t_o \in T$ and consider the process $(X_t)_{t \in T} := (\log \frac{dP_t}{dP_{t_o}})_{t \in T}$. If $\alpha \in A(T)$, $t_o \notin \alpha$, $\alpha_o := \alpha \cup \{t_o\}$ and $z \in S_{\alpha_o}$, then

$$\int \exp\Big(\sum_{t \in \alpha} z_t X_t\Big)\, dP_{t_o} = H(E_{\alpha_o})(z).$$

Let K be the kernel of E standardized at $t_o \in T$. Then we obtain

$$\log H(E_{\alpha_o})(z) = \frac{1}{2} \sum_{s,t \in \alpha_o} z_s z_t K(s,t) - \frac{1}{2} \sum_{t \in \alpha_o} z_t K(t,t) =$$

$$= \frac{1}{2} \sum_{s,t \in \alpha} z_s z_t K(s,t) - \frac{1}{2} \sum_{t \in \alpha} z_t K(t,t) .$$

Hence, by the Uniqueness theorem for Laplace transforms, $(X_t)_{t \in T}$ is a Gaussian process under P_{t_o} with mean $(-\frac{1}{2} K(t,t))_{t \in T}$ and covariance kernel K .

For the proof that (1) follows from (3) let $t_o \in T$ be such that $(X_t)_{t \in T} = (\log \frac{dP_t}{dP_{t_o}})_{t \in T}$ is a Gaussian process under P_{t_o} with covariance K . Homogeneity implies $P_{t_o}(\exp X_t) = 1$ and therefore $P_{t_o}(X_t) = -\frac{1}{2} K(t,t)$, $t \in T$. Computing the Laplace transforms of a Gaussian process with covariance K and mean $(-\frac{1}{2} K(t,t))_{t \in T}$ leads to Hellinger transforms of the form considered in Lemma (2.4). □

<u>(2.12) Corollary.</u> Suppose that $E = (\Omega, A, \{P_t : t \in T\})$ is a Gaussian experiment. Then for every $t_o \in T$ the kernel K standardized at t_o is the covariance structure of $(\log \frac{dP_t}{dP_{t_o}})_{t \in T}$ under P_{t_o} ; moreover,

$$P_{t_o}(\log \frac{dP_t}{dP_{t_o}}) = -\frac{1}{2} K(t,t), \quad t \in T .$$

In Strasser, 1984 a , it is shown how certain invariance conditions for Gaussian experiments can be put in terms of restrictions on the underlying class of Gaussian processes (see also Ex. (2.5) (2)).

3. Introduction to Poisson Experiments

Let $T \neq \emptyset$. If $E = (\Omega, A, \{P_t: t \in T\})$ is an experiment for the parameter space T and $c > 0$, we may consider the random experiment which consists in first selecting a sample size $n \in \mathbb{N}_0$ according to a Poisson variable with expectation c and then carrying out the n-fold direct product $(\Omega^n, A^n, \{P_t^n: t \in T\})$ of E. (Here Ω^0 consists of a single point.) This experiment can be described as follows. As sample space take the direct sum $(\bigoplus_{n=0}^{\infty} \Omega^n, \bigoplus_{n=0}^{\infty} A^n)$, the underlying set of probability measures is $\{\varepsilon(c\,P_t): t \in T\}$, the normalized exponential of any finite measure μ on A being defined by

$$\varepsilon(\mu) := e^{-\mu(\Omega)} \bigoplus_{n=0}^{\infty} \frac{\mu^n}{n!}.$$

In analogy to the definition of compound Poisson measures we have the following

<u>(3.1) Definition.</u> Let μ_t, $t \in T$, be finite measures on some measurable space (Ω, A). Then $(\bigoplus_{n=0}^{\infty} \Omega^n, \bigoplus_{n=0}^{\infty} A^n, \{\varepsilon(\mu_t): t \in T\})$ is called the <u>compound Poisson experiment with intensities μ_t, $t \in T$</u>.

<u>(3.2) Examples.</u> (Compound Poisson experiments)

(1) Let $\Omega = \{\omega\}$ be a singleton, $\mu_t := t\,\varepsilon_\omega$ and P_t the Poisson distribution with expectation $t \in T := (0, \infty)$. Then the compound Poisson experiment with intensities μ_t, $t \in T$, is obviously equivalent to $(\mathbb{N}_0, 2^{\mathbb{N}_0}, \{P_t: t \in T\})$.

(2) In Chap. III we shall see that the experiment consisting of the distributions of so-called Poisson processes with intensities μ_t, $t \in T$, is equivalent to the compound Poisson experiment with the same intensities.

In order to calculate the Hellinger transform of a compound Poisson experiment, we need maps $\psi_z: \bigcup_{\alpha \in A(T)} [0, \infty)^\alpha \longrightarrow (-\infty, 0]$, $z \in \bigcup_{\alpha \in A(T)} S_\alpha$, defined by

$$\psi_z(x) := \prod_{t \in \alpha} x_t^{z_t} - \sum_{t \in \alpha} z_t x_t, \quad x \in [0, \infty)^\alpha, \quad z \in S_\alpha, \quad \alpha \in A(T).$$

Note that this definition is independent of the choice of α.

Let $\mathcal{B}_\alpha$ denote the Borel-σ-field of S_α, $\alpha \in A(T)$.

<u>(3.3) Lemma.</u> Let E be a compound Poisson experiment with intensities μ_t, $t \in T$. Then

$$H(E_\alpha)(z) = \exp\Big(\int_{S_\alpha} \psi_z \, dM_\alpha \Big), \quad z \in S_\alpha$$

where

$$M_\alpha | \mathcal{B}_\alpha := \mathcal{L}\Big(\Big(\frac{d\mu_t}{d \sum_{s \in \alpha} \mu_s}\Big)_{t \in \alpha} \Big| \sum_{s \in \alpha} \mu_s \Big), \quad \alpha \in A(T).$$

<u>Proof.</u> Let $\alpha \in A(T)$ and $z \in S_\alpha$. Then

$$H(E_\alpha)(z) = \int_{S_\alpha} \prod_{t \in \alpha} \Big(\frac{d\varepsilon(\mu_t)}{d\varepsilon(\sum_{s \in \alpha} \mu_s)} \Big)^{z_t} d\varepsilon\Big(\sum_{s \in \alpha} \mu_s\Big) =$$

$$\prod_{t \in \alpha} e^{-z_t \|\mu_t\|} \sum_{n=0}^{\infty} \frac{1}{n!} \int \prod_{t \in \alpha} \Big(\frac{d\mu_t^n}{d(\sum_{s \in \alpha} \mu_s)^n} \Big)^{z_t} d\Big(\sum_{s \in \alpha} \mu_s\Big)^n =$$

$$\exp\Big(- \sum_{t \in \alpha} z_t \int_{S_\alpha} x_t \, M_\alpha(dx) \Big) \sum_{n=0}^{\infty} \frac{1}{n!} \Big(\int_{S_\alpha} \prod_{t \in \alpha} x_t^{z_t} \, M_\alpha(dx) \Big)^n. \qquad \square$$

Compound Poisson experiments need not be homogeneous. However, every compound Poisson experiment is at least pairwise imperfect.

From Lemma (3.3) we see that the equivalence class of any compound Poisson experiment is completely determined by the measures M_α, $\alpha \in A(T)$. Obviously, these measures are finite. In many situations, however, pairwise imperfect limit experiments are obtained the Hellinger transforms of which are of the form given in Lemma (3.3) with σ-finite measures M_α, $\alpha \in A(T)$. It is therefore desirable to fix a notion of Poisson experiments such that these cases are covered as well.

We need some technical prerequisites. For $t \in T$ let $p_t : [0,\infty)^T \longrightarrow [0,\infty)$ denote the projection onto the tth coordinate. Define $s_\alpha^2 := \sum_{t \in \alpha} (p_t - \frac{1}{|\alpha|})^2$ and $e_\alpha \in S_\alpha$ by $p_t(e_\alpha) := \frac{1}{|\alpha|}$, $t \in \alpha$, $\alpha \in A(T)$.

<u>(3.4) Lemma.</u> Let T be finite and M_T a measure on $\mathcal{B}_T$. Then there

is a pairwise imperfect experiment E for the parameter space T such that

$$H(E)(z) = \exp\Big(\int_{S_T} \psi_z \, dM_T\Big), \quad z \in S_T,$$

iff $\int_{S_T} s_T^2 \, dM_T < \infty$.

Proof. Suppose E exists. Then

$$\int_{S_T} s_T^2 \, dM_T \leq \frac{4}{|T|} \sum_{s \in T} \sum_{t \in T} \int_{S_T} (\sqrt{p_s} - \sqrt{p_t})^2 \, dM_T$$

$$= -\frac{8}{|T|} \sum_{s \in T} \sum_{t \in T} \log(1 - d^2(P_s, P_t)) < \infty .$$

Conversely, assume the integrability condition holds. For every $n \in \mathbb{N}$ define a finite measure

$$M_T^{(n)}: A \longmapsto M_T(A \cap \{s_T^2 > \tfrac{1}{n}\}), \quad A \in \mathcal{B}_T$$

and let $E^{(n)}$ be the compound Poisson experiment with intensities $p_t \cdot M_t^{(n)}$, $t \in T$. Then

$$H(E^{(n)})(z) = \exp\Big(\int_{\{s_T^2 > \frac{1}{n}\}} \psi_z \, dM_T\Big), \quad z \in S_T, \quad n \in \mathbb{N}.$$

Hence, $(E^{(n)})_{n \in \mathbb{N}}$ converges to an experiment E with Hellinger transforms given by

$$H(E)(z) = \exp\Big(\int_{S_T} \psi_z \, dM_T\Big), \quad z \in S_T .$$

To prove the pairwise imperfectness of E let $z \in S_T$ and define

$$Q_z: x \longmapsto \frac{|T|}{2} \Big(\sum_{s \in T} \sum_{t \in T} z_s z_t \big(x_s - \tfrac{1}{|T|}\big)\big(x_t - \tfrac{1}{|T|}\big) - \sum_{t \in T} z_t \big(x_t - \tfrac{1}{|T|}\big)^2 \Big),$$

$$x \in [0, \infty)^T .$$

By the Cauchy-Schwarz inequality Q_z is M_T-integrable. Note that $\psi_z(e_T) = Q_z(e_T) = 0$, $\frac{\partial}{\partial x_s} \psi_z(e_T) = \frac{\partial}{\partial x_s} Q_z(e_T) = 0$ $(s \in T)$ and

$$\frac{\partial^2}{\partial x_s \, \partial x_t} \psi_z(e_T) = \frac{\partial^2}{\partial x_s \, \partial x_t} Q_z(e_T) = |\alpha| z_s z_t - |\alpha| z_t \delta_{st}, \quad \{s,t\} \subset T$$

(here δ_{st} denotes Kronecker's Delta).

Now, a Taylor expansion of Q_z and ψ_z around e_T shows that for every $\varepsilon > 0$ there is a $\delta > 0$ such that $s_T^2(x) < \delta$ implies $|\psi_z(x) - Q_z(x)| < \varepsilon$.

Hence, ψ_z is M_T-integrable, too. In particular, $H(E_{\{s,t\}}) > 0$ for every $(s,t) \in T^2$. □

<u>(3.5) Definition.</u> Let $\alpha \in A(T)$. A measure M_α on $\mathcal{B}_\alpha$ is a <u>Lévy measure</u> if $M_\alpha\{e_\alpha\} = 0$ and s_α^2 is M_α-integrable.

The notion of Lévy measures is adapted from the theory of infinitely divisible distributions where similar concepts play an important rôle. If $\alpha \in A(T)$ and M_α is a measure on $\mathcal{B}_\alpha$ then M_α is a Lévy measure iff $M_\alpha\{e_\alpha\} = 0$ and $\int \psi_z \, dM_\alpha < \infty$, $z \in S_\alpha$. This was shown in the preceding proof. The restriction $M_\alpha\{e_\alpha\} = 0$ is imposed for convenience only; since $s_\alpha^2(e_\alpha) = \psi_z(e_\alpha) = 0$, $z \in S_\alpha$, it is without any influence.

<u>(3.6) Definition.</u> An experiment E for the parameter space T is a <u>Poisson experiment</u> if there are Lévy measures $M_\alpha | \mathcal{B}_\alpha$, $\alpha \in A(T)$, such that

$$H(E_\alpha)(z) = \exp\Big(\int_{S_\alpha} \psi_z \, dM_\alpha \Big), \quad z \in S_\alpha, \quad \alpha \in A(T).$$

This terminology will be justified in Part III. There it is shown that it coincides with the one introduced by LeCam, 1974: Every Poisson experiment is equivalent to an experiment consisting of distributions of generalized Poisson processes and vice versa.

<u>(3.7) Discussion.</u> Let E be a Poisson experiment for the parameter space T. E is pairwise imperfect. Since $H(E_\alpha)(z) = H(E_\beta)(z)$ if $\alpha \subset \beta \in A(T)$ and $z \in S_\alpha$, any system of Lévy measures $(M_\alpha)_{\alpha \in A(T)}$ giving the Hellinger transforms of E is <u>compatible</u> in the sense that

$$\int_{S_\alpha} \psi_z \, dM_\alpha = \int_{S_\beta} \psi_z \, dM_\beta, \quad \alpha \subset \beta \in A(T), \quad z \in S_\alpha.$$

Conversely, for every compatible system of Lévy measures $(M_{(\alpha)})_{\alpha \in A(T)}$ there is a projective system of experiments $(E_{(\alpha)})_{\alpha \in A(T)}$ satisfying

$$\int_{S_\alpha} \psi_z \, dM_{(\alpha)} = H(E_{(\alpha)})(z), \quad z \in S_\alpha, \quad \alpha \in A(T).$$

This follows from Lemma (3.4). The pertaining projective limit is a Poisson experiment with Lévy measures $M_{(\alpha)}$, $\alpha \in A(T)$.

Observe that the Lévy measures of any Poisson experiment are uniquely determined. This is due to the following uniqueness theorem for Lévy measures.

(3.8) Theorem. Let $\alpha \in A(T)$ and M^j, $j = 1,2$, be Lévy measures on $\mathcal{B}_\alpha$. If $\int \psi_z \, dM^1 = \int \psi_z \, dM^2$, $z \in S_\alpha$, then $M^1 = M^2$.

Proof. As a first step we prove that M^1 and M^2 coincide on $\overset{\circ}{S}_\alpha$. Let $H \subset \mathbb{R}^\alpha$ be the hyperplane orthogonal to e_α and $N :=$ $\{(z,h) \in \overset{\circ}{S}_\alpha \times H : z + h \in \overset{\circ}{S}_\alpha, \; z - h \in \overset{\circ}{S}_\alpha\}$. If $h = (h_t) \in H$ we denote

$$N_h := \{z \in \overset{\circ}{S}_\alpha : (z,h) \in N\},$$

$$\varphi_h : x \longmapsto 1 - \frac{1}{2} \prod_{t \in \alpha} x_t^{h_t} - \frac{1}{2} \prod_{t \in \alpha} x_t^{-h_t}, \quad (x_t) \in (0,\infty)^\alpha$$

and

$$\sigma_h : A \longmapsto \int_A \varphi_h \, d(M_1 - M_2), \quad A \in \overset{\circ}{S}_\alpha \cap \mathcal{B}_\alpha .$$

Since the maps φ_h vanish at e_α of second order, every σ_h is a bounded signed Borel measure on $\overset{\circ}{S}_\alpha$. Now, we fix any $h \in H$ such that $N_h \neq \emptyset$.

Let $s \in \alpha$ and $\ell := (\log \frac{p_t}{p_s})_{t \in \alpha \setminus \{s\}} |_{\overset{\circ}{S}_\alpha}$. Then

$$\begin{aligned} \int_{\ell(\overset{\circ}{S}_\alpha)} \exp\Big(\sum_{t \in \alpha \setminus \{s\}} z_t y_t\Big) \, L(\ell | p_s \cdot \sigma_h)(dy) &= \\ \int_{\overset{\circ}{S}_\alpha} \prod_{t \in \alpha} x_t^{z_t} \, \sigma_h(dx) &= \\ \int_{S_\alpha} \Big(\psi_z - \frac{1}{2}(\psi_{z-h} + \psi_{z+h})\Big) \, d(M^1 - M^2) &= 0, \quad z \in N_h . \end{aligned}$$

Since N_h is relatively open in $\overset{\circ}{S}_\alpha$, the Uniqueness theorem for the Laplace transform implies $L(\ell | p_s \cdot \sigma_h) = 0$ and hence $p_s \, \sigma_h |_{\overset{\circ}{S}_\alpha \cap \mathcal{B}_\alpha} = 0$. As $p_s |_{\overset{\circ}{S}_\alpha} > 0$, we obtain

$$\varphi_h \cdot M^1 |_{\overset{\circ}{S}_\alpha \cap \mathcal{B}_\alpha} = \varphi_h \cdot M^2 |_{\overset{\circ}{S}_\alpha \cap \mathcal{B}_\alpha} .$$

This renders $M^1 = M^2$ on $\overset{\circ}{S}_\alpha \cap \bigcup\{\varphi_h \neq 0 : h \in H, \; N_h \neq \emptyset\}$. To complete the first step note that

$$\varphi_h(x) = 0 \quad \text{iff} \quad \prod_{t \in \alpha} x_t^{h_t} = 1 ; \quad x \in \overset{\circ}{S}_\alpha, \quad h \in H$$

and that for every $x \in \overset{\circ}{S}_\alpha \setminus \{e_\alpha\}$ there is an $h \in H$ satisfying $N_h \neq \emptyset$

and $\prod_{t \in \alpha} x_t^{h_t} \neq 1$.

It remains to show that M^1 and M^2 coincide on $\partial S_\alpha := S_\alpha \setminus \mathring{S}_\alpha$. From the preceding step we obtain that

$$\int_{\partial S_\alpha} \sum_{t \in \alpha} z_t x_t M^1(dx) = \int_{\partial S_\alpha} \sum_{t \in \alpha} z_t x_t M^2(dx) , \quad z \in S_\alpha .$$

For reasons of continuity this relation remains valid even if $z \in \partial S_\alpha$. It follows that

$$\int_{\partial S_\alpha} \prod_{t \in \alpha} x_t^{z_t} M^1(dx) = \int_{\partial S_\alpha} \prod_{t \in \alpha} x_t^{z_t} M^2(dx) , \quad z \in S_\alpha .$$

Now, let $\beta \subset \alpha$ such that $|\beta| = |\alpha| - 1$. Since

$$\prod_{t \in \beta} x_t^{z_t} = 0 , \quad (x_t) \in \partial S_\alpha \setminus \mathring{S}_\beta , \quad z \in \mathring{S}_\beta$$

we obtain

$$\int_{\mathring{S}_\beta} \prod_{t \in \beta} x_t^{z_t} M^1(dx) = \int_{\mathring{S}_\beta} \prod_{t \in \beta} x_t^{z_t} M^2(dx) , \quad z \in S_\beta .$$

Another application of the Uniqueness theorem for the Laplace transform yields $M^1|_{\mathring{S}_\beta \cap \mathcal{B}_\alpha} = M^2|_{\mathring{S}_\beta \cap \mathcal{B}_\alpha}$ and therefore

$$\int_{\partial_1 S_\alpha} \prod_{t \in \alpha} x_t^{z_t} M^1(dx) = \int_{\partial_1 S_\alpha} \prod_{t \in \alpha} x_t^{z_t} M^2(dx), \quad z \in S_\alpha$$

where

$$\partial_1 S := \partial S_\alpha \setminus \cup \{\mathring{S}_\beta : \beta \subset \alpha , |\beta| = |\alpha| - 1\} .$$

Proceeding inductively, one obtains the assertion. □

We finish this section with a result characterizing the class of compound Poisson experiments within the class of all Poisson experiments.

(3.9) Theorem. A Poisson experiment is equivalent to a compound Poisson experiment iff each of its Lévy measures is finite.

Proof. The necessity of the condition follows from Lemma (3.3).

For the proof of the converse which runs similar to the construction of the projective limit of a compatible system of experiments (cf.

Siebert, 1979, Sec. 1) we need Choquet's theory of conical measures. Let E be a Poisson experiment for the parameter space T with bounded Lévy measures M_α, $\alpha \in A(T)$. Let $H(T)$ denote the smallest vector lattice of real-valued maps on $\mathbb{R}^T$ containing the projection maps p_t, $t \in T$. Note that the functions in $H(T)$ are positively homogeneous, each of which are only dependent on a finite number of coordinates. As the system (M_α) is compatible, we have

$$\int_{S_\alpha} \psi_z \, dM_\alpha = \int_{S_\alpha} \psi_z \, d\, L\Big(\big(\frac{p_t}{\sum_{s \in \alpha} p_s} \big)_{t \in \alpha} \,\Big|\, \sum_{s \in \alpha} p_s M_\beta \Big),$$

$$z \in S_\alpha, \quad \alpha \subset \beta \in A(T)$$

and therefore (cf. (3.8))

$$M_\alpha = L\Big(\big(\frac{p_t}{\sum_{s \in \alpha} p_s} \big)_{t \in \alpha} \,\Big|\, \sum_{s \in \alpha} p_s M_\beta \Big), \quad \alpha \subset \beta \in A(T).$$

Hence,

$$M: \varphi \longmapsto \int_{S_\alpha} \varphi \, dM_\alpha \quad \text{if} \quad \varphi \in H(T) \quad \text{and} \quad \varphi = \varphi \circ (p_t)_{t \in \alpha}$$

is a well-defined conical measure on $H(T)$.

Let μ be a Baire measure on $[0, \infty)^T$ such that

$$M\varphi = \int \varphi \, d\mu, \quad \varphi \in H(T)$$

(cf. Siebert, 1979, Sec. 1, Prop. 2), and consider the compound Poisson experiment G with intensities $p_t \mu$, $t \in T$. We claim that $E \sim G$. Let $\alpha \in A(T)$ and $z \in S_\alpha$. As ψ_z is continuous and concave on S_α, we have

$$\psi_z(y) = \inf \{ \sum_{t \in \alpha} x_t y_t : x \in D \}, \; y \in S_\alpha$$

where $D := \{ x \in \mathbb{R}^\alpha : \sum_{t \in \alpha} x_t u_t \geq \psi_z(u)$ for every $u \in S_\alpha \}$.

Now, let $\{x_i : i \in \mathbb{N}\}$ be a countable dense subset of D and φ_k be the support function of $\{x_1, \ldots, x_k\}$, $k \in \mathbb{N}$. Then (φ_k) decreases to ψ_z on S_α. This gives

$$\int_{S_\alpha} \psi_z \, dM_\alpha = \inf_{k \in \mathbb{N}} \int_{S_\alpha} \varphi_k \, dM_\alpha = \inf_{k \in \mathbb{N}} \int \varphi_k \, d\mu$$

$$= \inf_{k \in \mathbb{N}} \int_{S_\alpha} \varphi_k \, d\mu = \int_{S_\alpha} \psi_z \, d\mu,$$

since $L((p_t)_{t \in \alpha} | \mu)$ puts mass zero outside S_α. For the same reason the restriction of μ to B_α is the Lévy measure of G_α. □

4. Convergence of Poisson Experiments

In this section we shall characterize the weak convergence of Poisson experiments in terms of the convergence of the associated Lévy measures, thus obtaining a general representation of the Hellinger transforms of the possible limit experiments. Some of the following results are obviously known and used by LeCam, 1974 (Sec. 9), but not all of them are stated and proved there. The exposition will be given for nets instead of sequences, as the space of experiment-types is weakly compact (LeCam, 1974, p. 29) but in general not weakly sequentially compact. Moreover, nets will be needed in the third part.

Let $T \neq \emptyset$ and $E_\nu = (\Omega_\nu, A_\nu, \{P_{\nu t}: t \in T\})$, $\nu \in N$, be a net of experiments for the parameter space T.

<u>(4.1) Definition.</u> $(E_\nu)_{\nu \in N}$ is <u>bounded</u> if every weak accumulation point of $(E_\nu)_{\nu \in N}$ is pairwise imperfect.

<u>(4.2) Remark.</u> Since the space of experiment types for T is weakly compact (LeCam, 1974, p. 29) and weak convergence implies convergence of Hellinger distances, $(E_\nu)_{\nu \in N}$ is bounded iff

$$\overline{\lim_{\nu \in N}}\ d(P_{\nu s}, P_{\nu t}) < 1, \quad (s,t) \in T^2 .$$

In the case of nets of Poisson experiments boundedness may be phrased in terms of the associated Lévy measures.

<u>(4.3) Lemma.</u> Let E_ν, $\nu \in N$, be Poisson experiments. For every $\nu \in N$ let $(M_{\nu\alpha})_{\alpha \in A(T)}$ be the system of Lévy measures of E_ν. Then $(E_\nu)_{\nu \in N}$ is bounded iff

$$\overline{\lim_{\nu \in N}}\ \int s_\alpha^2 \, dM_{\nu\alpha} < \infty \quad \text{for every } \alpha \in A(T).$$

<u>Proof.</u> Similar to the proof of Lemma (3.4) it can be checked that

$$\overline{\lim_{\nu \in N}}\ \int s_\alpha^2 \, dM_{\nu\alpha} < \infty, \quad \alpha \in A(T),$$

iff the condition given in the preceding remark is fulfilled. □

Let us call a net of Lévy measures <u>bounded</u> if it satisfies the condition of Lemma (4.3).

Convention. For the following we fix $\alpha \in A(T)$ and denote the space of continuous real-valued maps on S_α by $C(S_\alpha)$.

The next definition is again adapted from the theory of infinitely divisible distributions.

(4.4) Definition. A net $(M_\nu)_{\nu \in N}$ of Lévy measures on $\mathcal{B}_\alpha$ is Lévy convergent if for every pair $(s,t) \in \alpha^2$ there is a bounded signed measure $K^{s,t}$ on $\mathcal{B}_\alpha$ such that

$$\lim_{\nu \in N} \int_{S_\alpha} f(p_s - \frac{1}{|\alpha|})(p_t - \frac{1}{|\alpha|})\, dM_\nu = \frac{1}{|\alpha|} \int f\, dK^{s,t}, \quad f \in C(S_\alpha).$$

In case of existence the system $(K^{s,t})_{s,t \in \alpha}$ is called the limit of $(M_\nu)_{\nu \in N}$.

We show that a bounded net of Poisson experiments converges weakly iff all associated nets of Lévy measures are Lévy convergent. The proof is divided into several steps.

(4.5) Lemma. Any bounded net (sequence) of Lévy measures on $\mathcal{B}_\alpha$ contains a Lévy convergent subnet (subsequence).

Proof. Let $(M_\nu)_{\nu \in N}$ be a bounded net of Lévy measures on S_α and $g_t = p_t - \frac{1}{|\alpha|}$, $t \in \alpha$. Then

$$\sup_{\nu \in N} \left| \int f\, g_s\, g_t\, dM_\nu \right| < \infty, \quad f \in C(S_\alpha), \quad (s,t) \in \alpha^2.$$

Hence, the sets $\{g_s g_t \cdot M_\nu : \nu \in N\}$, $(s,t) \in \alpha^2$, are relatively compact in the space of bounded signed measures on S_α, equipped with the vague topology, and $\prod_{(s,t) \in \alpha^2} \{g_s g_t M_\nu : \nu \in N\}$ is relatively compact in the corresponding product space. □

(4.6) Lemma. Let $(M_\nu)_{\nu \in N}$ be a Lévy convergent net of Lévy measures on $\mathcal{B}_\alpha$ and $(K^{s,t})_{s,t \in \alpha}$ be the limit. Then:

(1) $M(A) := \frac{1}{|\alpha|} \int_A s_\alpha^{-2} \, d \sum_{t \in \alpha} K^{t,t}$, $A \in \mathcal{B}_\alpha$, $e_\alpha \notin A$ defines a Lévy measure M on $\mathcal{B}_\alpha$, and (M_ν) converges to M , vaguely on $S_\alpha \setminus \{e_\alpha\}$.

(2) $K: (s,t) \longmapsto K^{s,t}\{e_\alpha\}$, $(s,t) \in \alpha^2$, is a positive semi-definite kernel.

(3) The limit $(K^{s,t})_{s,t \in \alpha}$ is completely determined by the pair (K, M).

Proof. Let $g_t := p_t - \frac{1}{|\alpha|}$, $t \in \alpha$.

(1) Since $\int s_\alpha^2 \, dM = \frac{1}{|\alpha|} \sum_{t \in \alpha} K^{t,t}(S_\alpha) < \infty$, M is a Lévy measure.

To prove vague convergence let $f \in C(S_\alpha)$ such that $e_\alpha \notin \operatorname{supp} f$. Then $f \, s_\alpha^{-2} \in C(S_\alpha)$, and hence

$$\lim_{\nu \in N} \int f \, dM_\nu = \frac{1}{|\alpha|} \sum_{t \in \alpha} \lim_{\nu \in N} |\alpha| \int_{S_\alpha} \frac{f}{s_\alpha^2} g_t^2 \, dM_\nu = \int_{S_\alpha} f \, dM .$$

(2) Let $f_n : S_\alpha \longrightarrow [0,1]$, $n \in \mathbb{N}$, be a sequence of continuous maps decreasing to $1_{\{e_\alpha\}}$. For $n \in \mathbb{N}$ and $\nu \in N$ let $\langle \cdot , \cdot \rangle_{n,\nu}$ denote the inner product in $L^2(f_n M_\nu)$. Then (2) follows from

$$K(s,t) = \lim_{n \to \infty} \lim_{\nu \in N} \langle g_s , g_t \rangle_{n,\nu} , \quad (s,t) \in \alpha^2 .$$

(3) Let $(s,t) \in \alpha^2$ and $(f_n)_{n \in \mathbb{N}}$ as above. Additionally, assume that $e_\alpha \notin \operatorname{supp}(1 - f_n)$, $n \in \mathbb{N}$. Then

$$\int f(1 - f_n) \, dK^{s,t} = |\alpha| \int f(1 - f_n) \, g_s \, g_t \, dM, \quad f \in C(S_\alpha), \; n \in \mathbb{N} ,$$

by (1), showing that

$$\int f \, dK^{s,t} = f(e_\alpha) K(s,t) + |\alpha| \lim_{n \to \infty} \int f(1 - f_n) g_s \, g_t \, dM, \quad f \in C(S_\alpha).$$

Thus, the linear form $f \longmapsto \int f \, dK^{s,t}$ on $C(S_\alpha)$ is uniquely determined by (K, M) . □

In the following we characterize the limit of a Lévy convergent net of Lévy measures by the pair constructed in Lemma (4.6).

(4.7) Theorem. Suppose $(M_\nu)_{\nu \in N}$ is a Lévy convergent net of Lévy measures on S_α with limit pair (K, M). Then

$$\lim_{\nu\in N} \int \psi_z \, dM_\nu = \frac{1}{2} \Big(\sum_{s\in\alpha} \sum_{t\in\alpha} z_s z_t K(s,t) - \sum_{t\in\alpha} z_t K(t,t)\Big) + \int \psi_z \, dM, \quad z\in S_\alpha .$$

<u>Proof.</u> Let $z \in S_T$, $g_t := p_t - \frac{1}{|\alpha|}$ $(t\in\alpha)$ and define

$$Q_z := \frac{|\alpha|}{2} \sum_{s\in\alpha} \sum_{t\in\alpha} z_s z_t g_s g_t - \sum_{t\in\alpha} z_t g_t^2 .$$

For every $\varepsilon > 0$ select a $\delta(\varepsilon) > 0$ such that $s_\alpha^2(x) < \delta$ implies $|Q_z(x) - \psi_z(x)| < \varepsilon$ (cf. the proof of (3.4)); w.l.g. assume $\delta(\varepsilon)$ is strictly decreasing to zero, as $\varepsilon \longrightarrow 0$. Choose $h_\varepsilon \in C(S_\alpha)$ satisfying

$$1_{\{s_\alpha^2 < \delta(\frac{\varepsilon}{2})\}} \leq h_\varepsilon \leq 1_{\{s_\alpha^2 \leq \delta(\varepsilon)\}}, \quad \varepsilon > 0 .$$

Then

$$\int \psi_z \, dM_\nu = \int Q_z h_\varepsilon \, dM_\nu + \int \psi_z (1-h_\varepsilon) \, dM_\nu + \sum_{t\in\alpha} \int \frac{\psi_z - Q_z}{s_\alpha^2} h_\varepsilon g_t^2 \, dM_\nu , \quad \nu \in N, \ \varepsilon > 0$$

where

$$\lim_{\nu\in N} \int Q_z h_\varepsilon \, dM_\nu = \frac{1}{2}\Big(\sum_{s\in\alpha} \sum_{t\in\alpha} z_s z_t \int h_\varepsilon \, dK^{s,t} - \sum_{t\in\alpha} z_t \int h_\varepsilon \, dK^{t,t}\Big), \ \varepsilon > 0 ,$$

$$\lim_{\nu\in N} \int \psi_z (1-h_\varepsilon) \, dM_\nu = \int \psi_z (1-h_\varepsilon) \, dM , \quad \varepsilon > 0 \qquad \text{and}$$

$$\overline{\lim_{\nu\in N}} \Big| \sum_{t\in\alpha} \int \frac{\psi_z - Q_z}{s_\alpha^2} h_\varepsilon g_t^2 \, dM_\nu \Big| \leq \frac{\varepsilon}{|\alpha|} \sum_{t\in\alpha} \int h_\varepsilon \, dK^{t,t} , \quad \varepsilon > 0 .$$

Hence, $(\int \psi_z \, dM_\nu)_{\nu\in N}$ is a bounded net, and every pertaining accumulation point c satisfies

$$\Big|c - \frac{1}{2}\Big(\sum_{s\in\alpha} \sum_{t\in\alpha} z_s z_t \int h_\varepsilon \, dK^{s,t} - \sum_{t\in\alpha} z_t \int h_\varepsilon \, dK^{s,t}\Big) - \int \psi_z (1-h_\varepsilon) dM\Big| \leq \frac{\varepsilon}{|\alpha|} \sum_{t\in\alpha} \int h_\varepsilon \, dK^{t,t} , \quad \varepsilon > 0 .$$

Now, the assertion follows from

$$\lim_{\varepsilon\to 0} \int \psi_z (1-h_\varepsilon) \, dM = \int \psi_z \, dM$$

and

$$\lim_{\varepsilon\to 0} \int h_\varepsilon \, dK^{s,t} = K(s,t), \quad (s,t) \in \alpha^2 . \qquad \square$$

<u>(4.8) Corollary.</u> Let E_ν, $\nu \in N$, be Poisson experiments, and suppose that $(E_\nu)_{\nu \in N}$ is bounded. Then $(E_\nu)_{\nu \in N}$ converges weakly iff the associated nets $(M_{\nu\beta})_{\nu \in N}$, $\beta \in A(T)$, of Lévy measures are Lévy convergent.

<u>Proof.</u> If each $(M_{\nu\beta})_{\nu \in N}$ is Lévy convergent, then $(E_\nu)_{\nu \in N}$ converges weakly according to the preceding theorem. Conversely, Lévy convergence of each of the nets $(M_{\nu\beta})_{\nu \in N}$ is necessary in view of Lemmas (4.3) and (4.5) and Theorems (3.8) and (4.7). □

In order to derive a representation of weak accumulation points of Poisson nets from Theorem (4.7), we require an extension of the uniqueness theorem for Lévy measures.

<u>(4.9) Theorem.</u> Let M^j be Lévy measures on $\mathcal{B}_\alpha$ and K_j positive semidefinite kernels on α^2, $j = 1,2$. For $z \in S_T$ let

$$L_j(z) := \frac{1}{2}\left(\sum_{s \in \alpha} \sum_{t \in \alpha} z_s z_t K(s,t) - \sum_{t \in \alpha} z_t K(t,t)\right), \quad j = 1,2 .$$

If $L_1(z) + \int \psi_z \, dM^1 = L_2(z) + \int \psi_z \, dM^2$, $z \in S_T$, then $L_1 = L_2$ and $M^1 = M^2$.

<u>Proof.</u> Keep the notations of the proof of Theorem (3.8). The proof that M^1 and M^2 coincide on $\overset{\circ}{S}_\alpha$ can be copied from there, the only difference being that

$$\int_{\overset{\circ}{S}_\alpha} \prod_{t \in \alpha} x_t^{z_t} \, \sigma_h(dx) = \frac{1}{2} \sum_{s \in \alpha} \sum_{t \in \alpha} h_s h_t (K_1(s,t) - K_2(s,t)), \; h \in H, \; z \in N_h .$$

But this does not cause problems, since it still implies that σ_h is concentrated on e_α for every $h \in H$.

To show $L_1 = L_2$ we choose a $C > 0$ such that N_h is non-void if $h \in H$ and $0 < \max_{t \in \alpha} |h_t| \leq C$. Let $\{s \neq t\} \subset \alpha$ and define $h_r = 0$ if $r \notin \{s,t\}$ and $h_s = C = -h_t$. Then $(h_r) \in H$, $N_{(h_r)} \neq \emptyset$ and

$$\sum_{q \in \alpha} \sum_{r \in \alpha} h_q h_r (K_1(q,r) - K_r(q,r)) = 0$$

by the first step. Thus,

$$K_1(s,s) - K_2(s,s) + K_1(t,t) - K_2(t,t) - 2 K_1(s,t) + 2 K_2(s,t) = 0 .$$

By Lemma (2.6) K_1 and K_2 define equivalent Gaussian experiments, which yields $L_1 = L_2$.

Now, an application of Theorem (3.8) completes the proof. □

The following corollary which is crucial for the development of Section 5 can also be found in LeCam, 1974 (Lemma 3, p. 77).

(4.10) Corollary. If $(E_\nu)_{\nu \in N}$ is a bounded net of Poisson experiments, then every weak accumulation point of (E_ν) is equivalent to a direct product of a Gaussian and a Poisson experiment. Both factors are unique up to equivalence.

Proof. Let E be a weak accumulation point of $(E_\nu)_{\nu \in N}$. From Theorem (4.7) it follows that every finite subexperiment E_α, $\alpha \in A(T)$ is equivalent to a direct product of a Gaussian and a Poisson experiment. By Theorem (4.9) both, the Gaussian and the Poisson factors, form projective systems of experiments. Taking the respective projective limits yields the existence of the factorization. Its uniqueness is again a consequence of Theorem (4.9). □

5. Convergence of Triangular Arrays

Let $T \neq \emptyset$ be an arbitrary set. Following LeCam, 1974, an experiment E for the parameter space T will be called infinitely divisible if for every $n \in \mathbb{N}$ there exists an nth root E_n of E, i.e. an experiment E_n for the parameter space T such that $E_n^n \sim E$. Note that in case of existence roots of experiments are uniquely determined up to equivalence.

We show that the class of pairwise imperfect and infinitely divisible experiments coincides with the class of weak limits of infinitesimal and bounded triangular arrays of experiments, the technical terms being defined analogously to the theory of infinitely divisible distributions (for precise definitions see below).

Let us first recall some basic facts about infinitely divisible experiments. The following easy lemma is sometimes useful.

(5.1) Lemma. An experiment E for the parameter space T is infinitely divisible iff for every $n \in \mathbb{N}$ and every $\alpha \in A(T)$ $H(E_\alpha)^{1/n}$ is the Hellinger transform of an experiment for the parameter space α.

Proof. Apparently, the condition is necessary. For the proof of sufficiency let $n \in \mathbb{N}$ and $E_{n,(\alpha)}$ be experiments such that

$$H(E_\alpha)^{1/n} = H(E_{n,(\alpha)}), \quad \alpha \in A(T).$$

Then the system $E_{n,(\alpha)}$, $\alpha \in A(T)$ is projective, and the projective limit is an nth root of E. □

(5.2) Remark. The following statements are evident either from the definition of infinite divisibility or from the preceding lemma.

(1) The direct product of a finite number of infinitely divisible experiments for the same parameter space is infinitely divisible.

(2) Every weak limit of a net of infinitely divisible experiments is infinitely divisible.

<u>(5.3) Example.</u> Gaussian experiments, Poisson experiments, and direct products of both are infinitely divisible as follows from Lemma (5.1) and Remark (5.2) (1).

In the remaining part of this chapter we deal with the following situation. Let $(k_n)_{n \in \mathbb{N}} \subset \mathbb{N}$ be a sequence increasing to infinity. Consider experiments $E_{ni} = (\Omega_{ni}, A_{ni}, \{P_{nit}: t \in T\})$, $1 \leq i \leq k_n$, $n \in \mathbb{N}$. The double sequence $(E_{ni})_{1 \leq i \leq k_n,\, n \in \mathbb{N}}$ is a triangular array of experiments. We investigate the asymptotic behaviour of the product experiments

$$E_n = \prod_{i=1}^{k_n} E_{ni} = \left(\prod_{i=1}^{k_n} \Omega_{ni},\ \bigotimes_{i=1}^{k_n} A_{ni},\ \{\bigotimes_{i=1}^{k_n} P_{nit}: t \in T\}\right),\quad n \in \mathbb{N}.$$

For convenience we denote $P_{nt} = \bigotimes_{i=1}^{k_n} P_{nit}$, $n \in \mathbb{N}$, $t \in T$.

<u>(5.4) Definition.</u>

(1) The array $(E_{ni})_{1 \leq i \leq k_n,\, n \in \mathbb{N}}$ is <u>bounded</u> if the sequence $(E_n)_{n \in \mathbb{N}}$ is bounded.

(2) $(E_{ni})_{1 \leq i \leq k_n,\, n \in \mathbb{N}}$ is <u>infinitesimal</u> if for every pair $(s,t) \in T^2$

$$\lim_{n\to\infty} \max_{1 \leq i \leq k_n} d^2(P_{nis}, P_{nit}) = 0.$$

Intuitively speaking, the assumption of infinitesimality guarantees that the influence of any single factor within a row of the array becomes uniformly negligibly small, as the row number tends to infinity. Before turning to an example, we give alternative characterizations of boundedness and infinitesimality. For this, we need the following technicalities.

Let $\|\cdot\|$ denote the variational norm of bounded signed measures.

<u>(5.5) Lemma.</u> Let $E = (\Omega, A, \{P_1,\ldots,P_m\})$ be a finite experiment. Then

$$1 - H(E)(z) \leq (m-1) \max_{1 \leq j \leq m} \|P_m - P_j\|,\quad z \in S_{\{1,\ldots,m\}}.$$

Proof. Let $z \in S_{\{1,\dots,m\}}$, $\min_{1 \le k \le m} z_k > 0$, $\nu := \sum_{k=1}^{m} P_k$, $f_k \in \frac{dP_k}{d\nu}$ and

$$A_k := \{f_{m-k+1} \le \prod_{\ell=1}^{m-k} f_\ell^{\frac{z_\ell}{1-z_m-\dots-z_{m-k+1}}}\}, \quad 1 \le k \le m. \text{ Then}$$

$$H(E)(z) = \int_{A_1} f_m \left(\frac{\prod_{\ell=1}^{m-1} f_\ell^{\frac{z_\ell}{1-z_m}}}{f_m} \right)^{1-z_m} d\nu +$$

$$\int_{\Omega \setminus A_1} \prod_{\ell=1}^{m-1} f_\ell^{\frac{z_\ell}{1-z_m}} \left(\frac{f_m}{\prod_{\ell=1}^{m-1} f_\ell^{\frac{z_\ell}{1-z_m}}} \right)^{z_m} d\nu \ge$$

$$\int_{A_1} f_m \, d\nu + \int_{\Omega \setminus A_1} \prod_{\ell=1}^{m-1} f_\ell^{\frac{z_\ell}{1-z_m}} d\nu .$$

Proceeding inductively, we obtain

$$1 - H(E)(z) \le 1 - \sum_{j=1}^{m} \int_{A_j \cap \bigcap_{\ell=1}^{j-1} (\Omega \setminus A_\ell)} f_{m-j+1} \, d\nu =$$

$$\sum_{j=2}^{m} \int_{A_j \cap \bigcap_{\ell=1}^{j-1} (\Omega \setminus A_\ell)} (f_m - f_{m-j+1}) \, d\nu \le$$

$$\sum_{j=2}^{m} \| P_m - P_{m-j+1} \| ,$$

and hence the assertion. □

(5.6) Remark. Let $0 \le a_j \le 1$ $(1 \le j \le n)$. For the following recall

$$\prod_{i=1}^{n} a_i \le \exp\left(- \sum_{i=1}^{n} (1 - a_i)\right) \le \prod_{i=1}^{n} a_i + \frac{1}{2} \max_{1 \le i \le n} (1 - a_i) .$$

(5.7) Lemma.

(1) If the triangular array $(E_{ni})_{1 \le i \le k_n}$, $n \in \mathbb{N}$ is bounded, then

$$\overline{\lim_{n\to\infty}} \sum_{i=1}^{k_n} d^2(P_{nis}, P_{nit}) < \infty \quad \text{for every pair } (s,t) \in T^2 .$$

For infinitesimal arrays the converse is also true.

(2) The array $(E_{ni})_{1\le i\le k_n,\ n\in\mathbb{N}}$ is infinitesimal iff for every $\alpha \in A(T)$ and every $z \in S_\alpha$

$$\lim_{n\to\infty} \max_{1\le i\le k_n} (1 - H(E_{ni,\alpha})(z)) = 0 .$$

Proof. Assertion (1) follows from

$$1 - \exp(- \sum_{i=1}^{k_n} d^2(P_{nis}, P_{nit})) \le d^2(P_{ns}, P_{nt}) \le$$

$$1 - \exp(- \sum_{i=1}^{k_n} d^2(P_{nis}, P_{nit})) + \frac{1}{2} \max_{1\le i\le k_n} d^2(P_{nis}, P_{nit}) ,$$

$$(s,t) \in T^2, \ n \in \mathbb{N}$$

which itself is a consequence of the preceding remark.

Ad (2). Taking $\alpha = \{s,t\}$ for arbitrary pairs in T^2 shows the sufficiency of the condition. For the proof of its necessity let $\alpha \in A(T)$, $s \in \alpha$ and $z \in S_\alpha$. Then, by the preceding lemma,

$$\max_{1\le i\le k_n} (1 - H(E_{ni,\alpha}))(z) \le$$

$$|\alpha| \max_{1\le i\le k_n} \max_{t\in\alpha} \|P_{nis} - P_{nit}\| \le$$

$$2\,|\alpha| \max_{1\le i\le k_n} \max_{t\in\alpha} d(P_{nis}, P_{nit}), \quad n \in \mathbb{N} ,$$

which tends to zero if the triangular array is infinitesimal. □

As a consequence of Lemma (5.7) (1) boundedness implies infinitesimality in case each row of the array consists of identical factors.

(5.8) Example. Now, we take up again the situation described already in the first section. Let Q be a probability measure on $(\mathbb{R}^1, \mathcal{B}(\mathbb{R}^1))$, Q_s its translate under $s \in \mathbb{R}^1$ and $c_{ni} = (c_{ni\ell}) \in \mathbb{R}^k$, $1 \le i \le k_n$. We consider the regression problems $E_n = \bigotimes_{i=1}^{k_n} E_{ni}$,

$$E_{ni} := (\mathbb{R}^1, \mathcal{B}(\mathbb{R}^1), \{Q_{<c_{ni},t>} : t \in \mathbb{R}^k\}), \quad 1 \leq i \leq k_n ,$$

generated by the design matrices $C_n := (c_{n1}, \ldots, c_{nk_n})^T$ $(n \in \mathbb{N})$ and look for conditions on (C_n) under which the array (E_{ni}) has only non-trivial weak accumulation points.

For this, the rate of convergence of $d(Q_0, Q_s) \longrightarrow 0$ (as $s \longrightarrow 0$) turns out to be of importance. We assume the existence of constants $a \geq 1$, $b \geq 0$ and $c > 0$ such that $a = 1$ implies $b = 0$ and

(1) $$\lim_{s \to 0} \frac{d(Q_0, Q_s)}{|s|^a \, (\log \frac{1}{|s|})^b} = c .$$

This covers most reasonable cases. If the translation parameter experiment $(\mathbb{R}^1, \mathcal{B}(\mathbb{R}^1), \{Q_s : s \in \mathbb{R}^1\})$ is L^2-differentiable with a finite and strictly positive Fisher's information I, then (1) holds with $a = 2$, $b = 0$ and $c = \frac{I}{8}$. Some other typical examples are given below.

Under (1) the array (E_{ni}) is infinitesimal and bounded iff

(2) $$\lim_{n \to \infty} \max_{1 \leq i \leq k_n} |c_{ni\ell}| = 0 \quad \text{and}$$

$$\overline{\lim_{n \to \infty}} \sum_{i=1}^{k_n} |c_{ni\ell}|^a \, (\log \frac{1}{|c_{ni\ell}|})^b < \infty , \quad 1 \leq \ell \leq k .$$

To see that infinitesimality and boundedness imply (2) note that for some $\delta > 0$

$$x \longmapsto x^a (\log \tfrac{1}{x})^b , \quad x \in [0, \delta)$$

is a convex and strictly increasing function which varies regularly at zero. For the reverse implication note also that

$$s \longmapsto d(Q_0, Q_s) , \quad s \in \mathbb{R}^1$$

is bounded away from zero. Due to the first inequality in the proof of Lemma (5.7) (1) every weak accumulation point of (E_n) is identifiable (i.e. its parametrization is one-to-one) iff

(3) $$\underline{\lim_{n \to \infty}} \sum_{i=1}^{k_n} d^2(Q_0, Q_{<c_{ni},t>}) > 0 , \quad t \in \mathbb{R}^k .$$

In general, this condition cannot be put in a form similar to (2). Summarizing this discussion we can say that, in the presence of (1), Conditions (2) and (3) ensure that every sequence of subexperiments of E_n, $n \in \mathbb{N}$, has only non-trivial weak accumulation points.

Now, let us consider the particular case of a k-sample problem in location. In order to find an appropriate choice of the sequence of design matrices C_n, $n \in \mathbb{N}$, we pursue the idea of localization explained already in the introduction. For $n \in \mathbb{N}$ let $\{k_n(\ell): 1 \le \ell \le k\}$ be such that $\sum_{\ell=1}^{k_n} k_n(\ell) = k_n$; assume that the nth experiment consists of $k_n(\ell)$ observations from each sample ℓ, $1 \le \ell \le k$. On a "non-local level" these experiments are given by the row-wise products of the array

$$(\mathbb{R}^1, \mathcal{B}(\mathbb{R}^1), \{Q_{<\bar{c}_{ni},t>} : t \in \mathbb{R}^k\}, \quad 1 \le i \le k_n, \quad n \in \mathbb{N},$$

where

$$\bar{c}_{ni} = \left(1_{(\sum_{j=1}^{\ell-1} k_n(j), \sum_{j=1}^{\ell} k_n(j)]}(i)\right)_{1 \le \ell \le k}.$$

Since this parametrization violates Condition (2), it leads to a totally informative weak limit experiment. Similar to the one-sample location parameter case we look for a reparametrizing sequence $\delta_n \downarrow 0$ such that (2) and (3) hold for

$$c_{ni} := \delta_n \cdot \bar{c}_{ni}, \quad 1 \le i \le k_n, \quad n \in \mathbb{N}.$$

Employing (1) it is not hard to see that the sequence

$$\delta_n := k_n^{-\frac{1}{a}} (\log k_n)^{-\frac{b}{a}}$$

will do.

For further reference we mention the following specific examples.

(a) Let Q denote the Γ-distribution with variance $\delta > 0$:

$$\frac{dQ}{d\lambda^1}(x) := \Gamma(\delta)^{-1} x^{\delta-1} e^{-x}, \quad x > 0.$$

Then (1) holds. For $\delta > 2$ we have $a = 2$, $b = 0$ and $c = \frac{1}{8(\delta-2)}$ (the translation parameter family generated by Q is differentiable in quadratic mean with finite Fisher's Information $\frac{1}{\delta-2}$); for $\delta = 2$ we have $a = 2$, $b = 1$ and $c = 2$ and for $\delta < 2$ $a = \delta$, $b = 0$ and $c = \Gamma(\delta)^{-1} \left(\frac{1}{\delta} + \int_1^\infty (y^{\frac{\delta-1}{2}} - (y-1)^{\frac{\delta-1}{2}})^2 dy\right)$, as lengthy but elementary calculations show (cf. also Ibragimov and Has'minskii, 1981, Sec. II 5 and VI, Theorem (1.1)).

(b) Let Q denote the truncated standard normal distribution with truncation points ± 1, i.e.

$$\frac{dQ}{d\lambda^1}(x) := \frac{1}{2\Phi(1) - 1} \varphi(x) \cdot 1_{(-1,+1)}(x), \quad x \in \mathbb{R}^1,$$

where φ denotes the standard normal density and Φ the respective distribution function. Then

$$d(Q, Q_s) = 1 - e^{-\frac{s^2}{2}} \frac{2\,\Phi(1 - \frac{|s|}{2}) - 1}{2\,\Phi(1) - 1}, \quad s \in \mathbb{R}^1$$

showing that (1) holds with $a = 1$, $b = 0$ and $c = \frac{\varphi(1)}{2\,\Phi(1) - 1}$.

(c) If Q stands for the Pareto-distribution with (fixed) scale parameter $\gamma > 0$; then

$$\frac{dQ}{d\lambda^1}(x) := \frac{\gamma}{x+1}, \quad x \geq 1 \qquad \text{and}$$

$$d^2(Q_o, Q_s) = 1 - \gamma\left(\frac{2}{s}\right)^{\gamma} \int_{1+\frac{2}{s}}^{\infty} \frac{dy}{(y^2-1)^{\frac{1+\gamma}{2}}}, \quad s > 0.$$

Now, routine estimations show that (4) holds with $a = 1$, $b = 0$ and $c = \frac{\gamma}{2}$.

For further examples cf. e.g. Becker, 1983, Strasser, 1984 a and various other sources.

Now, we show that to each infinitesimal triangular array one can assign a sequence of compound Poisson experiments G_n, $n \in \mathbb{N}$, which is weakly asymptotically equivalent to the sequence of row-wise products of the array. This reduces the asymptotic theory of triangular arrays to the convergence theory for Poisson experiments which has already been presented in Sec. 4. For $n \in \mathbb{N}$ let G_n, $n \in \mathbb{N}$, be the compound Poisson experiment with intensities $\bigoplus_{i=1}^{k_n} P_{nit}$, $t \in T$.

(5.9) Theorem. If $(E_{ni})_{1 \leq i \leq k_n,\, n \in \mathbb{N}}$ is infinitesimal, then $(E_n)_{n \in \mathbb{N}}$ and $(G_n)_{n \in \mathbb{N}}$ are weakly asymptotically equivalent. Precisely, for every $\alpha \in A(T)$

$$\lim_{n\to\infty} (H(E_{n\alpha}) - H(G_{n\alpha})) = 0 \text{ on } S_\alpha.$$

Proof. Let $\alpha \in A(T)$ and $z \in S_\alpha$. Employ Lemma (3.3) to obtain

$$H(G_{n\alpha})(z) = \exp\left[\sum_{i=1}^{k_n} (H(E_{ni,\alpha})(z) - 1)\right].$$

Now, the application of Remark (5.6) with $a_i = H(E_{ni,\alpha})$, $1 \leq i \leq k_n$, yields

$$|H(G_{n\alpha})(z) - H(E_{n\alpha})(z)| \leq \max_{1 \leq i \leq k_n} (1 - H(E_{ni,\alpha})(z)).$$

Taking into consideration Lemma (5.7) the proof is complete. □

In view of this theorem $(G_n)_{n \in \mathbb{N}}$ will be called the <u>sequence of Poisson experiments accompanying the triangular array</u>.

<u>(5.10) Corollary.</u> Suppose $(E_{ni})_{1 \leq i \leq k_n,\ n \in \mathbb{N}}$ is infinitesimal and bounded. Then every weak accumulation point of $(E_n)_{n \in \mathbb{N}}$ is equivalent to a direct product of a Gaussian and a Poisson experiment.

<u>Proof.</u> Combine Corollary (4.10) and Theorem (5.9). □

We summarize the main results of this section in the following theorem which supplements and partly restates LeCam's characterization of infinitely divisible experiments (LeCam, 1974, Prop. 2, p. 79).

<u>(5.11) Theorem.</u> The following classes of experiment-types coincide:

(1) The pairwise imperfect infinitely divisible experiments.

(2) The direct products of a Gaussian and a Poisson factor.

(3) The weak accumulation points of bounded sequences of Poisson experiments.

(4) The weak accumulation points of infinitesimal and bounded triangular arrays of experiments.

<u>Proof.</u> Let $\mathcal{E}_i$, $1 \leq i \leq 4$, be the above described classes of experiment-types. Then $\mathcal{E}_1 \supset \mathcal{E}_2$ by Example (5.3), $\mathcal{E}_2 \supset \mathcal{E}_3$ because of the preceding corollary, $\mathcal{E}_3 \supset \mathcal{E}_4$ on account of Theorem (5.9) and $\mathcal{E}_4 \supset \mathcal{E}_1$, since the roots of any pairwise imperfect and infinitely divisible experiment form an infinitesimal and bounded array. □

<u>(5.12) Remark.</u> Let E be a pairwise imperfect infinitely divisible experiment for the parameter space T. By Theorems (4.7), (4.9) and (5.11) there are a covariance kernel K on T and a compatible system of Lévy measures M_α on $\mathcal{B}_\alpha$, $\alpha \in A(T)$, such that

$$H(E_\alpha)(z) = \exp\left[\frac{1}{2}\left(\sum_{s,t\in\alpha} z_s z_t K(s,t) - \sum_{t\in\alpha} z_t K(t,t)\right) + \int_{S_\alpha} \psi_z \, dM_\alpha\right],$$
$$\alpha \in A(T),\ z \in S_\alpha .$$

(M_α) is unique, and K is uniquely determined up to equivalence. This is the celebrated <u>Lévy-Khintchine representation</u> for the Hellinger transforms of infinitely divisible experiments. For a derivation from a Lévy-Khintchine representation of standard measures cf. Part II, Sec. 10.

6. Identification of Limit Experiments

Consider the same situation as in the preceding section, i.e. a triangular array $E_{ni} = (\Omega_{ni}, A_{ni}, \{P_{nit}: t \in T\})$, $1 \leq i \leq k_n$, $n \in \mathbb{N}$, of experiments for some parameter space T and the appertaining sequence $(E_n)_{n \in \mathbb{N}}$ of product experiments. Let $(M_{n\alpha})_{\alpha \in A(T)}$, $n \in \mathbb{N}$, be the systems of Lévy measures which define the sequence of Poisson experiments accompanying $(E_{ni})_{1 \leq i \leq k_n,\, n \in \mathbb{N}}$. Throughout this whole section we shall assume that $(E_{ni})_{1 \leq i \leq k_n,\, n \in \mathbb{N}}$ is infinitesimal and bounded. We give conditions under which all weak accumulation points of $(E_n)_{n \in \mathbb{N}}$ are either Gaussian or Poisson experiments.

<u>(6.1) Definition.</u> The array $(E_{ni})_{1 \leq i \leq k_n,\, n \in \mathbb{N}}$ is <u>Gaussian</u> if $(E_n)_{n \in \mathbb{N}}$ has only Gaussian accumulation points.

<u>(6.2) Theorem.</u> $(E_{ni})_{1 \leq i \leq k_n,\, n \in \mathbb{N}}$ is Gaussian iff

(G) $\lim\limits_{n \to \infty} M_{n\alpha}\{s_\alpha^2 > \varepsilon\} = 0$ for every $\alpha \in A(T)$ and every $\varepsilon > 0$.

<u>Proof.</u> Suppose $(E_{ni})_{1 \leq i \leq k_n,\, n \in \mathbb{N}}$ is Gaussian. Let $\alpha \in A(T)$ and $\mathbb{N}_1 \subset \mathbb{N}$ be an arbitrary subsequence. Extract a subsequence $\mathbb{N}_2 \subset \mathbb{N}_1$ such that $(E_{n\alpha})_{n \in \mathbb{N}_2}$ converges weakly to a Gaussian experiment for the parameter space α. In view of Lemma (2.4) and the results of Section 4 there exists a positive semi-definite kernel K_α on α^2 such that $(K_\alpha, 0)$ is the limit pair of $(M_{n\alpha})_{n \in \mathbb{N}_2}$. Hence, $(M_{n\alpha})_{n \in \mathbb{N}_2}$ converges to 0, vaguely on $S_\alpha \setminus \{e_\alpha\}$, showing the validity of (G) by means of a standard contradiction argument.

Now, assume that (G) holds. Let E be an accumulation point of $(E_n)_{n \in \mathbb{N}}$, $\alpha \in A(T)$ and M_α the Lévy measure of the Poisson factor of E_α. Further, let $(E_{n\alpha})_{n \in \mathbb{N}_1}$ be a subsequence of $(E_{n\alpha})_{n \in \mathbb{N}}$ converging to E_α, weakly. Since $(M_{n\alpha})_{n \in \mathbb{N}_1}$ converges to M_α, vaguely on $S_\alpha \setminus \{e_\alpha\}$, condition (G) implies $M_\alpha = 0$. Thus, E_α is equivalent to its Gaussian factor. □

Condition (G) which characterizes asymptotic Gaussian behaviour can be simplified considerably. This is mainly due to the fact that for every $n \in \mathbb{N}$ and every $\alpha \in A(T)$

$$M_{n\alpha}(A) = \sum_{t \in \alpha} \sum_{i=1}^{k_n} P_{nit}\{ (\frac{dP_{nis}}{d \sum_{r \in \alpha} P_{nir}})_{s \in \alpha} \in A \}$$

if $A \in \mathcal{B}_\alpha$ and $e_\alpha \notin A$ (cf. Lemma (3.3)).

<u>(6.3) Theorem.</u> The array $(E_{ni})_{1 \le i \le k_n,\, n \in \mathbb{N}}$ is Gaussian iff

$$\lim_{n \to \infty} \sum_{i=1}^{k_n} P_{nis}\{ |\frac{dP_{nit}}{dP_{nis}} - 1| > \varepsilon\} = 0 \quad \text{for every } \varepsilon > 0 \text{ and } (s,t) \in T^2 .$$

<u>Proof.</u> Assume that $(E_{ni})_{1 \le i \le k_n,\, n \in \mathbb{N}}$ is Gaussian, and let $(s,t) \in T^2$. Applying condition (G) with $\alpha = \{s,t\}$ yields

$$\lim_{n \to \infty} \sum_{i=1}^{k_n} P_{nis}\{ |\frac{dP_{nit}}{d(P_{nis} + P_{nit})} - \frac{1}{2}| > \varepsilon\} = 0 , \quad \varepsilon > 0$$

which gives the desired relation.

Conversely, suppose that the above-stated condition holds. Let $\alpha \in A(T)$, $n \in \mathbb{N}$ and $\varepsilon > 0$. Then (G) follows from

$$M_{n\alpha}\{s_\alpha^{\,2} > \varepsilon\} \le$$

$$\sum_{s \in \alpha} \sum_{r \in \alpha} \sum_{i=1}^{k_n} P_{nis}\{ |\frac{dP_{nir}}{d \sum_{t \in \alpha} P_{nit}} - \frac{1}{|\alpha|}| > \frac{\sqrt{\varepsilon}}{|\alpha|} \} \le$$

$$\sum_{s \in \alpha} \sum_{r \in \alpha} \sum_{t \in \alpha} \sum_{i=1}^{k_n} P_{nis}\{ |\frac{dP_{nit}}{dP_{nis}} - 1| > \frac{\sqrt{\varepsilon}}{|\alpha|}\} .$$ □

The condition of Theorem (6.3) is closely related to the Oosterhoff-van Zwet criteria for asymptotic normality of log-likelihood ratios (see Oosterhoff, van Zwet, 1979, Theorem 2). Since it does only rely on the likelihood ratios of all binary subexperiments of the E_{ni}, it is fairly easy to handle.

Gaussian experiments have the merit that they can be identified from their binary subexperiments. Similarly, the asymptotic behaviour of Gaussian arrays is completely determined by their "binary subarrays", as can be seen from the following corollary.

(6.4) Corollary. Assume that $(E_{ni})_{1 \le i \le k_n,\, n \in \mathbb{N}}$ is Gaussian. Then $(E_n)_{n \in \mathbb{N}}$ converges weakly iff $\left(\sum_{i=1}^{k_n} d^2(P_{nis}, P_{nit})\right)_{n \in \mathbb{N}}$ converges for every pair $(s,t) \in T^2$.

In the case of convergence to some experiment $E = (\Omega, A, \{P_t\colon t \in T\})$, E is Gaussian and $\lim_{n\to\infty} \sum_{i=1}^{k_n} d^2(P_{nis}, P_{nit}) = -\log(1 - d^2(P_s, P_t))$, $(s,t) \in T^2$.

Proof. If $E = (\Omega, A, \{P_t\colon t \in T\})$ is any accumulation point of $(E_n)_{n \in \mathbb{N}}$, $\alpha \in A(T)$ and $(E_{n\alpha})_{n \in \mathbb{N}_1}$ is a subsequence of $(E_{n\alpha})_{n \in \mathbb{N}}$ converging to E_α, then

$$\lim_{n \in \mathbb{N}_1} \sum_{i=1}^{k_n} d^2(P_{nis}, P_{nit}) = -\log(1 - d^2(P_s, P_t)), \quad (s,t) \in \alpha^2$$

on account of the inequalities given in the proof of Lemma (5.7) (1). Taking into consideration Theorem (2.7) and self-evident fact that the projective limit of Gaussian experiments is again Gaussian, then completes the proof. □

(6.5) Remark. Assume that $(E_{ni})_{1 \le i \le k_n,\, n \in \mathbb{N}}$ is a convergent Gaussian array and that the limits

$$a(s,t) := \lim_{n\to\infty} \sum_{i=1}^{k_n} d^2(P_{nis}, P_{nit}), \quad (s,t) \in T^2,$$

are given. Denote the limit of $(E_n)_{n \in \mathbb{N}}$ by $E = (\Omega, A, \{P_t\colon t \in T\})$. Let $t_o \in T$. Then the kernel K of E standardized at t_o is given by

$$K(s,t) = 4(a(s,t_o) + a(t,t_o) - a(s,t)), \quad (s,t) \in T^2,$$

as can be seen from the above Corollary and Discussion (2.9). Moreover,

$$P_{t_o}\left(\log \frac{dP_t}{dP_{t_o}}\right) = -4\, a(t,t_o), \quad t \in T$$

(cf. Corollary (2.12)).

(6.6) Remark. For results extending (6.3) and (6.4) cf. Becker, 1983, Satz (2.2.2) and Korollar (2.2.5). There it is shown that in Theorem (6.3) it is also sufficient to establish the condition for every $t \in T$ with s arbitrary but fixed. Applications and examples are given in

Becker, 1983, Chap. 4. Additionally, we remark that the i.i.d. location parameter array of Γ-distributions with scale parameter 2,

$$E_{ni} = (\mathbb{R}^1, \mathcal{B}(\mathbb{R}^1), \{Q_{\frac{t}{\sqrt{n \log n}}} : t \in \mathbb{R}^1\}), \quad 1 \le i \le n, \quad n \in \mathbb{N},$$

locally reparametrized around 0 (cf. Ex. (5.8) (1)) is Gaussian, as can easily be checked by means of Theorem (6.3). Moreover, since

$$\lim_{n\to\infty} n \cdot d^2(Q_0, Q_{\frac{t}{\sqrt{n \log n}}}) = t^2, \quad t \in \mathbb{R}^1,$$

the above remark implies that $(E_n)_n$ converges to a Gaussian experiment whose kernel standardized at 0 is given by $(s,t) \longmapsto 8\, s \cdot t$, $(s,t) \in \mathbb{R}^2$. Hence, (E_n) is LAN with "covariance" 8 (cf. e.g. Becker, 1983, (1.3.4)).

Now, we turn to the consideration of Poisson convergence.

<u>(6.7) Definition.</u> $(E_{ni})_{1 \le i \le k_n}$ is a <u>Poisson array</u> if each of its accumulation points is a Poisson experiment.

<u>(6.8) Theorem.</u> $(E_{ni})_{1 \le i \le k_n,\, n \in \mathbb{N}}$ is a Poisson array iff

(P) $$\lim_{\varepsilon\to 0} \overline{\lim_{n\to\infty}} \int_{\{s_\alpha{}^2 < \varepsilon\}} s_\alpha{}^2 \, dM_{n\alpha} = 0 \text{ for every } \alpha \in A(T).$$

<u>Proof.</u> To begin with let (E_{ni}) be a Poisson array, $\alpha \in A(T)$ and $\mathbb{N}_1 \subset \mathbb{N}$ be an arbitrary subsequence. Select a subsequence $\mathbb{N}_2 \subset \mathbb{N}_1$ such that $(E_{n\alpha})_{n \in \mathbb{N}_2}$ converges weakly to a Poisson experiment whose Lévy measure will be denoted by M_α. Then $(0, M_\alpha)$ is the limit pair of $(M_{n\alpha})_{n \in \mathbb{N}_2}$; hence,

$$\lim_{n \in \mathbb{N}_2} \int_{S_\alpha} f(p_s - \tfrac{1}{|\alpha|})(p_t - \tfrac{1}{|\alpha|})\, dM_{n\alpha} = \int_{S_\alpha} f(p_s - \tfrac{1}{|\alpha|})(p_t - \tfrac{1}{|\alpha|})\, dM_\alpha, \quad \{s,t\} \subset \alpha, \quad f \in C(S_\alpha)$$

(cf. the proof of Lemma (4.6)). Plainly, this implies (P).

Conversely, suppose that (P) is valid. Let E be a weak accumulation point of $(E_n)_{n \in \mathbb{N}}$, $\alpha \in A(T)$ and $(E_{n\alpha})_{n \in \mathbb{N}_1}$ be a subsequence of $(E_{n\alpha})$ converging to E. Because of Corollary (4.8) and Theorem (5.9)

$(M_{n\alpha})_{n\in\mathbb{N}_1}$ is Lévy convergent. Let (K_α, M_α) be the limit pair. Then

$$\lim_{n\in\mathbb{N}_1} \int f(p_s - \frac{1}{|\alpha|})(p_t - \frac{1}{|\alpha|})dM_{n\alpha} = \int f(p_s - \frac{1}{|\alpha|})(p_t - \frac{1}{|\alpha|})dM_\alpha + \frac{f(e_\alpha)}{|\alpha|} \cdot K_\alpha(s,t), \quad (s,t)\in\alpha^2, \ f\in C(S_\alpha).$$

Together with condition (P) and a standard integrability-argument this yields $K_\alpha(s,t) = 0$, $(s,t)\in\alpha^2$.

On the other hand K_α is a covariance kernel of the Gaussian factor of E_α (cf. Theorem (4.7)). □

Similar as for Gaussian arrays, Condition (P) can be phrased in terms of the likelihood ratios of the experiments E_{ni}, $1\le i\le k_n$, $n\in\mathbb{N}$.

<u>(6.9) Theorem.</u> $(E_{ni})_{1\le i\le k_n,\, n\in\mathbb{N}}$ is a Poisson array iff

$$\lim_{\varepsilon\to 0}\ \overline{\lim_{n\to\infty}}\ \sum_{i=1}^{k_n} \int_{\{|\frac{dP_{nis}}{dP_{nit}} - 1| < \varepsilon\}} |\frac{dP_{nis}}{dP_{nit}} - 1|^2 \, dP_{nit} = 0$$

for every pair $(s,t)\in T^2$.

<u>Proof.</u> From the remark preceding Theorem (6.3) we obtain that (P) is equivalent to

$$(P') \quad \lim_{\varepsilon\to 0}\ \overline{\lim_{n\to\infty}} \int_{\{\sum_{q\in\alpha} |\frac{dP_{niq}}{d\sum_{r\in\alpha} P_{nir}} - \frac{1}{|\alpha|}| < \varepsilon\}} \left(\frac{dP_{nis}}{d\sum_{r\in\alpha} P_{nir}} - \frac{1}{|\alpha|}\right)^2 dP_{nit} = 0$$

for every $\alpha\in A(T)$ and every pair $(s,t)\in\alpha^2$.

Suppose $(E_{ni})_{1\le i\le k_n,\, n\in\mathbb{N}}$ is a Poisson array, and let $(s,t)\in T^2$. Specifying condition (P') for $\alpha = \{s,t\}$ we obtain

$$\lim_{\varepsilon\to 0}\ \overline{\lim_{n\to\infty}}\ \sum_{i=1}^{k_n} \int_{\{|\frac{dP_{nis}}{d(P_{nis}+P_{nit})} - \frac{1}{2}| < \varepsilon\}} \left(\frac{dP_{nis}}{d(P_{nis}+P_{nit})} - \frac{1}{2}\right)^2 dP_{nit} = 0$$

which yields the desired relation.

To prove the converse let $\alpha\in A(T)$, $(s,t)\in\alpha^2$, $\varepsilon>0$, $n\in\mathbb{N}$ and $1\le i\le k_n$.

Then elementary calculations show that

$$\int\limits_{\{\sum_{q\in\alpha} |\frac{dP_{niq}}{d\sum_{r\in\alpha} P_{nir}} - \frac{1}{|\alpha|}| < \varepsilon\}} \left(\frac{dP_{nis}}{d\sum_{r\in\alpha} P_{nir}} - \frac{1}{|\alpha|}\right)^2 dP_{nit} \leq$$

$$\leq \frac{1}{|\alpha|} \int\limits_{\bigcup_{q\in\alpha}\bigcup_{r\in\alpha}\{|\frac{dP_{niq}}{dP_{nir}} - 1| < \frac{4\varepsilon}{|\alpha|}\}} \sum_{u\in\alpha} \left(\frac{\frac{dP_{niu}}{dP_{nit}}}{\frac{dP_{nis}}{dP_{nit}}} - 1\right)^2 \frac{dP_{nis}}{dP_{nit}} \, dP_{nit} \leq$$

$$\leq \sum_{q\in\alpha} \int\limits_{\{|\frac{dP_{niq}}{dP_{nir}} - 1| < 4\varepsilon\}} \left(\frac{dP_{niq}}{dP_{nit}} - 1\right)^2 dP_{nit} \, .$$

Hence, (P') follows from the condition stated in the theorem. □

Different to the Gaussian case, it is not possible to detect a Poisson experiment from its binary subexperiments, in general. Hence, without additional assumptions, for Poisson arrays results parallel to (6.4) and (6.5) cannot be expected to hold. This was one of the basic motivations for introducing the concept of experiments with independent increments (cf. the last contribution in this volume).

Before determining those "mixed cases" in which only compound Poisson factors can occur, we give some examples.

<u>(6.10) Examples.</u> Let us continue Examples (5.8). For simplicity we shall restrict ourselves to the consideration of localized one-sample location parameter families, i.e. of arrays

$$E_{ni} = (\mathbb{R}^1, \mathcal{B}(\mathbb{R}^1), \{Q_{\delta_n t}: t \in \mathbb{R}^1\}), \quad 1 \leq i \leq n, \quad n \in \mathbb{N}$$

with Q and (δ_n) as in (5.8). In view of Theorem (6.9) and the translation invariance of the E_{ni} these arrays will be Poisson iff

$$\text{(P'')} \qquad \lim_{\varepsilon\to 0} \overline{\lim_{n\to\infty}} \, n \int\limits_{\{|\frac{dQ_{\delta_n t}}{dQ_o} - 1| < \varepsilon\}} \left(\frac{dQ_{\delta_n t}}{dQ_o} - 1\right)^2 dP_o = 0, \quad t \in \mathbb{R}^1 .$$

(a) Consider the family of shifted Γ-distributions with fixed variance $\delta > 0$. As noted above, $(E_n)_n$ is LAN in case $\delta = 2$. For $\delta > 2$ the same is true, as follows from Hâjek's conditions for local asymptotic normality (cf. the appendix of Hâjek, 1972). Now, let $0 < \delta < 2$. Then it is not hard to verify (P"). Hence, (E_{ni}) is a Poisson array. Moreover,

$$\lim_{n\to\infty} n \cdot d^2(P_{n^{-\delta}s}, P_{n^{-\delta}t}) = \text{const}(\delta) \cdot |s-t|^{\delta}, \quad (s,t) \in \mathbb{R}^2 .$$

In view of the above remarks this does not prove convergence of (E_n). This may be done e.g. by considering the asymptotic behaviour of the Hellinger transforms. Re-examining the proof of Theorem (5.9) it can be seen that (E_n) converges weakly to some limit experiment E iff

$$\lim_{n\to\infty} \exp\left[\sum_{i=1}^{k_n} (H(E_{ni})(z) - 1)\right] = H(E_\alpha)(z), \quad z \in S_\alpha, \quad \alpha \in A(T).$$

For $1 < \delta < 2$ we obtain that $(E_n)_n$ converges to a translation invariant Poisson experiment E with Hellinger transforms

$$H(E_\alpha)(z) = \exp\Big(-\int_{t_k}^{\infty} \prod_{j=1}^{k} (x - t_j)^{(\delta-1)z_j}\, dx\Big),$$
$$\alpha = \{t_1 < \ldots < t_k\} \in A(\mathbb{R}^1), \quad z \in S_\alpha .$$

If $0 < \delta < 1$ a similar result holds.

To consider the case $\delta = 1$ we let F_σ denote the location parameter family of one-sided exponential distributions with variance $\sigma > 0$: $F_\sigma := (\mathbb{R}^1, \mathcal{B}(\mathbb{R}^1), \{P_t^{(\sigma)}: t \in \mathbb{R}^1\})$, where

$$\frac{dP_0^{(\sigma)}}{d\lambda^1}(x) := \frac{1}{\sigma}\, e^{-\frac{x}{\sigma}} \cdot 1_{[0,\infty)}(x), \quad x \in \mathbb{R}^1 .$$

Then

$$(1) \quad H(F_{\sigma\alpha})(z) = \exp\Big[\frac{1}{\sigma}\Big(\sum_{i=1}^{k} z_i t_i - \max_{\{i: z_i > 0\}} t_i\Big),$$
$$\alpha = \{t_1, \ldots, t_k\} \in A(\mathbb{R}^1), \quad z \in S_\alpha .$$

Hence, for $\delta = 1$, $E_n \sim F_1$, $n \in \mathbb{N}$. In particular, F_σ is a Poisson experiment for every $\sigma > 0$.

(b) In the case of renormalized standard normal distributions with densities truncated at ± 1, (E_{ni}) is a Poisson array, too. Denoting $\sigma := \frac{\varphi(1)}{2\,\Phi(1) - 1}$, we obtain

$$\lim_{n\to\infty} H(E_{n\alpha})(z) = \exp\Big[\sigma\Big(\min_{\{i: z_i > 0\}} t_i - \max_{\{i: z_i > 0\}} t_i\Big)\Big], \quad z \in S_\alpha,$$
$$\alpha = \{t_1, \ldots, t_k\} \in A(\mathbb{R}^1) .$$

Thus, (1) implies

$$E_n \longrightarrow F_\sigma \circledast F_{-\sigma}, \text{ weakly}$$

where $F_{-\sigma}$ denotes the reflection of F_σ at the origin.

(c) If Q stands for the Pareto distribution with (fixed) scale parameter $\gamma > 0$, then it is again easy to verify condition (P"). Hence, the appertaining array has only Poisson accumulation points. The proof of convergence requires more effort. Actually,

$$E_n \longrightarrow F_{\frac{1}{\gamma}}, \text{ weakly},$$

as can be demonstrated employing the theory of experiments with independent increments (cf. Ex. 19.5).

(6.11) Theorem. The following assertions are equivalent:

(1) The Poisson factor of every weak accumulation point of (E_{ni}) is equivalent to a compound Poisson experiment.

(2) $\lim\limits_{\varepsilon\to 0} \overline{\lim\limits_{n\to\infty}} M_{n\alpha}\{s_\alpha^2 > \varepsilon\} < \infty$ for every $\alpha \in A(T)$.

(3) $\lim\limits_{\varepsilon\to 0} \overline{\lim\limits_{n\to\infty}} \sum\limits_{i=1}^{k_n} P_{nis}\{|\frac{dP_{nit}}{dP_{nis}} - 1| > \varepsilon\} < \infty$ for every pair $(s,t) \in T^2$.

Proof. Let $(M_\alpha)_{\alpha \in A(T)}$ be the system of Lévy measures of some accumulation point of $(E_{ni})_{1 \le i \le k_n,\, n \in \mathbb{N}}$. Then

$$M_\alpha(S_\alpha) = \lim_{\varepsilon\to 0} M_\alpha\{s_\alpha^2 > \varepsilon\} = \lim_{\varepsilon\to 0} \overline{\lim_{m\to\infty}} M_{n_m\alpha}\{s_\alpha^2 > \varepsilon\}$$

if $\alpha \in A(T)$ and $(n_m)_{m\in\mathbb{N}}$ is such that $(E_{n_m\alpha})_{m\in\mathbb{N}}$ converges weakly to the respective subexperiment of the accumulation point under consideration. Thus, the equivalence of (1) and (2) follows from Theorem (3.9). The equivalence of (2) and (3) can be obtained entirely similar to the proof of Theorem (6.3). □

II. THE LÉVY-KHINTCHINE FORMULA FOR INFINITELY DIVISIBLE STATISTICAL EXPERIMENTS

Arnold Janssen

7. Preliminaries

Let Θ be a non-void parameter set. If E is an experiment for Θ, let $\dot{E}$ denote the equivalence class of E. We write $\mathcal{E}(\Theta)$ for the set of all equivalence classes of experiments for the parameter space Θ. If Θ is finite, then the classes in $\mathcal{E}(\Theta)$ can be described by standard measures on the unit simplex $S_\Theta := \{z \in [0,\infty)^\Theta \,/\, \Sigma z_\theta = 1\}$, which is a compact space in the induced topology. Recall that for finite Θ the measure

$$\mu := \Big(\sum_{\theta\in\Theta} P_\theta^{\,T}\Big) \quad \text{where} \quad T := \Big(\frac{dP_\sigma}{d \sum_{\theta\in\Theta} P_\theta}\Big)_{\sigma\in\Theta}$$

which belongs to $M_b(S_\Theta)$ is called the <u>standard measure</u> of the experiment $E = (X, \mathcal{A}, (P_\theta)_{\theta\in\Theta})$; by definition $G := (S_\Theta, \mathcal{B}(S_\Theta), ((P_\theta^{\,T})_{\theta\in\Theta}))$ is the <u>standard experiment</u> of E.

<u>(7.1) Remark.</u> For finite Θ the following statements hold (see Blackwell, 1951, Torgersen, 1970, p. 36 and LeCam, 1972).

(a) $E \sim G$.

(b) Two experiments for Θ are equivalent iff the corresponding standard measures coincide.

(c) A measure $\mu \in M_b(S_\Theta)$ is a standard measure iff $\int p_\theta \, d\mu = 1$ for every $\theta \in \Theta$. The standard experiment is then given by $(S_\Theta, \mathcal{B}(S_\Theta), (p_\theta\,\mu)_{\theta\in\Theta})$.

(d) A sequence of experiments (E_n) for Θ converges to E weakly iff

(i) the corresponding sequence (μ_n) of standard measures converges weakly to the standard measure μ of E, or - equivalently - iff

(ii) $p_i \mu_n \longrightarrow p_i \mu$, weakly, $i = 1, \ldots, k$.

Let $J \subset \Theta$ be finite and $\emptyset \neq I \subset J$. Then the mapping

$$\varphi_{I,J}: S_J \setminus \{(z_j)_{j \in J} / \sum_{i \in I} z_i = 0\} \longrightarrow S_I$$

defined by

$$\varphi_{I,J}((z_j)) := \left(\left(\frac{z_i}{\sum_{k \in I} z_k} \right)_{i \in I} \right)$$

is continuous. The proof of the next lemma is obvious.

(7.2) Lemma. Let J be finite, $I \subset J$ and $E = (S_J, \mathcal{B}(S_J), (P_j^J)_{j \in J})$, $F = (S_I, \mathcal{B}(S_I), (P_i^I)_{i \in I})$ be standard experiments. Then the following assertions are equivalent:

(a) $E_I \sim F$,

(b) $P_i^I = (P_i^J)^{\varphi_{I,J}}$ for every $i \in I$,

(c) $\sum_{i \in I} P_i^I = \left(\sum_{i \in I} P_i^J \right)^{\varphi_{I,J}}$.

The next definition is basic.

(7.3) Definition. Let E, E_t, $t > 0$, be experiments for Θ. Then E is called infinitely divisible if for every $n \in \mathbb{N}$ there exists an experiment E_n satisfying $(E_n)^n \sim E$. Let us call a family $(E_t)_{t > 0}$ a continuous semigroup of experiments if $E_t \otimes E_s \sim E_{t+s}$ for all $t, s > 0$ and $t \longmapsto (E_t)^{\cdot}$ is weakly continuous on $(0, \infty)$.

Since the definition depends only on the classes of the underlying experiments we shall speak of infinitely divisible classes and semigroups of classes. Let $\mathcal{EI}(\Theta)$ denote the set of classes of infinitely divisible experiments in $\mathcal{E}(\Theta)$.

Now, we want to describe the standard experiment of an experiment in case some of the underlying measures are mutually orthogonal.

(7.4) Lemma. Let $(S_\Theta, \mathcal{B}(S_\Theta), (\mu_\theta)_{\theta \in \Theta})$ be the standard experiment and μ the standard measure of an experiment $E = (X, \mathcal{A}, (P_\theta)_{\theta \in \Theta})$ for a finite space Θ. Suppose there is a partition $\Theta = \bigcup_{j=1}^{r} \Theta_j$ into r non-void parts such that P_i and P_s are mutually orthogonal, $P_i \perp P_s$, if i and s do not lie in some common set Θ_j. Let $(S_{\Theta_j}, \mathcal{B}(S_{\Theta_j}), (\nu_\theta^j)_{\theta \in \Theta_j})$ be the standard experiments of E_{Θ_j} and ν^j their standard measures. We define the injections $m_j: S_{\Theta_j} \longrightarrow S_\Theta$ by $p_\theta^\Theta(m_j(z)) := z_\theta$ if $\theta \in \Theta_j$ and $z = (z_\theta)_{\theta \in \Theta_j}$ and $p_\theta^\Theta(m_j(z)) := 0$ otherwise. Then

(i) $\mu_\theta = (\nu_\theta^j)^{m_j}$ for every $\theta \in \Theta_j$, $j = 1, \dots, r$.

(ii) $\mu = \sum_{j=1}^{r} (\nu^j)^{m_j}$.

<u>Proof.</u> If s and i lie in different sets Θ_j, then $p_s \cdot p_i = 0$ μ almost everywhere (a.e.). Hence, for every $\theta \in \Theta_j$ the measure μ_θ is concentrated on the set $A := \{z \in S_\Theta : z_\theta \neq 0,\ z_i = 0$ for all $i \notin \Theta_j\}$. By Lemma (7.2)

$$\nu_\theta^j = \mu_\theta^{\varphi_{\Theta_j,\Theta}}$$

follows. Now, we remark that the restriction of $m_j \circ \varphi_{\Theta_j,\Theta}$ to A is the identity. □

8. Infinitely Divisible Probability Measures

In this section we recall certain results and definitions for infinitely divisible probability measures on $\mathbb{R}^n$. Basic results about the Lévy-Khintchine formula can be found in the paper of Courrège, 1964; in a more general situation compare with Heyer, 1977.

<u>(8.1) Definition.</u> A probability measure $\mu \in M_1(\mathbb{R}^n)$ is called <u>infinitely divisible</u> if for every m there exists a mth root $\mu_m \in M_1(\mathbb{R}^n)$: $(\mu_m)^{*m} = \mu$.
A family $(\mu_t)_{t>0}$ in $M_1(\mathbb{R}^n)$ is called a <u>continuous convolution semigroup</u> if $\mu_t * \mu_s = \mu_{t+s}$ for all $t,s>0$ and $t \longmapsto \mu_t$ is weakly continuous on $(0,\infty)$.

<u>(8.2) Theorem.</u> Every infinitely divisible $\mu \in M_1(\mathbb{R}^n)$ is embeddable into a continuous convolution semigroup $(\mu_t)_{t>0}$ in $M_1(\mathbb{R}^n)$ such that $\mu = \mu_1$.

<u>(8.3) Discussion (Generating Functionals).</u> Let $C^2_{lok}(\mathbb{R}^n)$ denote the space of real-valued, continuous, bounded functions on $\mathbb{R}^n$ which are locally twice differentiable with continuous derivatives in some neighbourhood of 0. Then a continuous convolution semigroup (μ_t) is uniquely determined by its <u>generating functional</u>

(i) $$A(f) := \lim_{t\to 0} \frac{\int f\,d\mu_t - f(0)}{t} \quad \text{which exists on } C^2_{lok}(\mathbb{R}^n).$$

A admits a <u>Lévy-Khintchine representation</u>

(ii) $$A(f) = \langle \nabla f(0), b\rangle + \sum_{i,j=1}^{n} a_{ij} \frac{\partial}{\partial x_i}\frac{\partial}{\partial x_j} f(0) + \int_{\mathbb{R}^n \setminus \{0\}} \left(f(x) - f(0) - \frac{\langle \nabla f(0), x\rangle}{1+\|x\|^2}\right) d\eta(x).$$

In this formula

(iii) $b \in \mathbb{R}^n$,

(iv) $(a_{i,j})_{i,j=1,\dots,n}$ is a positive semi-definite real matrix *),

(v) η is a positive Radon measure on $\mathbb{R}^n \setminus \{0\}$ which may be un-

---

*) Positive semi-definite matrices are always meant to be symmetric.

bounded. η is defined by $\lim_{t \to 0} \frac{1}{t} \int f \, d\mu_t = \int f \, d\eta$ for $f \in C_{oo}(\mathbb{R}^n \setminus \{0\})$; it is called the <u>Lévy measure</u> of the convolution semigroup and fulfills

(vi) $\eta(\{y: \|y\| \geq 1\}) < \infty$ and

(vii) $\int_{\{y: \|y\| < 1\}} \|y\|^2 \, d\eta < \infty$.

In general, every Radon measure satisfying (vi) and (vii) will be called a <u>Lévy measure</u>. Conversely, if $(b, (a_{ij})_{i,j}, \eta)$ is a triplet satisfying (iii), (iv), (vi) and (vii), then there is a unique continuour convolution semigroup in $M_1(\mathbb{R}^n)$ having the generating functional A.

Let us recall some properties of continuous convolution semigroups $(\mu_t)_{t>0}$. First we note that μ_t converges weakly to ε_o as $t \longrightarrow 0$. If $A(f) = \langle \nabla f(0), b \rangle$ then $\mu_t = \varepsilon_{tb}$, $t \geq 0$; if η vanishes we obtain Gaussian measures, (8.5). Suppose that η is finite and $A(f) = \int_{\mathbb{R}^n \setminus \{0\}} (f(x) - f(0)) d\eta(x)$. Then A defines a <u>Poisson semigroup</u> $E(t\eta) := \exp(-t\|\eta\|) \sum_{n=0}^{\infty} \frac{t^n \eta^{*n}}{n!}$, $t \geq 0$ $(\eta^{*o} := \varepsilon_o)$. (Sometimes $E(t\eta)$ is said to be a <u>compound Poisson measure</u>).

(viii) Suppose that X is locally compact and consider a fixed point $x_o \in X$. A linear functional $A: V \longrightarrow \mathbb{R}$ on a linear subspace V of $C_b(X)$ is called

(a) <u>almost positive</u>, if for every $f \in V$, satisfying $f \geq f(x_o) = 0$, the relation $A(f) \geq 0$ holds;

(b) <u>tight</u>, if for each $\varepsilon > 0$ there exists a compact set $K \subset X$ such that for all $f \in V$, satisfying $f|_K = 0$ and $|f| \leq 1$, the assertion $|A(f)| \leq \varepsilon$ holds.

(ix) Suppose $X = \mathbb{R}^n$, $x_o = 0$, $V = C^2_{lok}(\mathbb{R}^n)$ and let A be a real functional on V. The following assertions are equivalent:

(1) A is the generating functional of a continuous convolution semigroup on $\mathbb{R}^n$,

(2) A is almost positive,
A is tight,
$A(1) = 0$.

Compare with Heyer, 1977, 4.4.18 and von Waldenfels, 1965. It should be mentioned that Courrège, 1964, deals only with a function space which is smaller than $C^2_{lok}(\mathbb{R}^n)$. But it is well-known that the results of (8.3) carry over.

(8.4) Remark. Applying (8.3) it is easy to compute the Lévy-Khintchine representation for the Fourier transforms $\hat{\mu}_t$. We have

(i) $\hat{\mu}_t(y) = \exp(t\,\Psi(y))$, where

(ii) $$\Psi(y) = i\langle y,b\rangle - \sum_{i,j=1}^{n} a_{ij}\,y_i\,y_j + \int_{\mathbb{R}^n\setminus\{0\}} \left(\exp(iyx) - 1 - \frac{i\langle y,x\rangle}{1+\|x\|^2}\right) d\eta(x).$$

(8.5) Definition. An infinitely divisible measure μ (continuous convolution semigroup $(\mu_t)_{t>0}$) is called a Gaussian measure (Gaussian semigroup) if its Lévy measure vanishes.

The family of Gaussian measures coincides with the set of distributions of $\mathbb{R}^n$-valued normally distributed random variables (including point measures).

(8.6) Remark. Let $C^2_{lok,A}(\mathbb{R}^n)$ denote the set of real-valued, continuous functions f on $\mathbb{R}^n$ (possibly unbounded) such that f is twice continuously differentiable in some neighbourhood of 0 and $\int_{\|x\|>1} |f|\,d\eta < \infty$. Then A has a canonical extension to $C^2_{lok,A}(\mathbb{R}^n)$.

(8.7) Lemma. Suppose that $\mu \in M_1(\mathbb{R}^n)$ is infinitely divisible. Let $(\mu_t)_{t>0}$ denote the corresponding continuous convolution semigroup such that $\mu = \mu_1$ and take A according to (8.3).
If $g\colon x \longmapsto \exp(\langle x,a\rangle)$, $x \in \mathbb{R}^n$ ($a = (a_i) \in \mathbb{R}^n$ fixed), then

(a) $\int g\,d\mu < \infty$ iff $\int_{\{\|x\|>1\}} g\,d\eta < \infty$.

(b) $\int g\,d\mu = 1$ iff $g \in C^2_{lok,A}(\mathbb{R}^n)$ and $A(g) = 0$.

(c) Suppose that the conditions of (b) hold. Then $\int g\,d\mu_t = 1$ for every $t>0$, and $(g\cdot\mu_t)_{t>0}$ is a continuous convolution semigroup having the generating functional $f \longmapsto A(gf)$ on $C^2_{lok}(\mathbb{R}^n)$.

<u>Proof.</u> By $K_r(0) \subset \mathbb{R}^n$ we denote the closed sphere with radius $r > 0$ centered at the origin. If η is a finite Lévy measure, then $E(\eta)$ is by definition the compound Poisson distribution generated by η. Fubini's theorem shows for $\alpha, \beta \in M_b(\mathbb{R}^n)$:

(1) $\int g \, d\alpha * \beta = (\int g \, d\alpha)(\int g \, d\beta)$.

(a) Consider the decomposition $\eta = \eta_1 + \eta_2$ of the Lévy measure η defined by $\eta_1 = \eta|_{K_1(0)}$ and $\eta_2 = \eta - \eta_1$. Then $\eta = \nu * E(\eta_2)$ is the convolution product of some infinitely divisible measure ν and $E(\eta_2)$. We remark that $f(x) := \exp(\sum_{i=1}^{n} |a_i x_i|)$ is submultiplicative (i.e. $f(x+y) \leq f(x) f(y)$). Now, a well-known result shows $\int f \, d\nu < \infty$ (and hence $\int g \, d\nu < \infty$) since ν has the Lévy measure η_1, see Kruglov, 1974, T. 3 and Kruglov, 1970. Paying regard to (1) we obtain:

$$\int g \, d \, E(\eta_2) < \infty \quad \text{iff} \quad \int_{\{\|x\| > 1\}} g \, d\eta_2 < \infty .$$

This is obvious.

(b) Suppose $\int g \, d\mu = 1$. Then (1) and Fubini's theorem imply $(g \cdot \mu_{1/n})^{*n} = g \cdot \mu$.
Hence, $g \cdot \mu$ is infinitely divisible. For every $f \in C_{oo}(\mathbb{R}^n) \cap C^2_{lok}(\mathbb{R}^n)$ the generating functional A_1 of $g \cdot \mu$ can be computed from

$$A_1(f) = \lim_{n \to \infty} n(\int f \, g \, d\mu_{1/n} - f(0)) = A(fg).$$

Let $f_m \in C_{oo}(\mathbb{R}^n)$ be a function such that $0 \leq f_m \leq 1$, $f_m|_{K_m(0)} = 1$ and $f_m|_{CK_{m+1}(0)} = 0$.

In view of (a) the monotone convergence theorem yields

$$A(g) = \lim_{m \to \infty} A(f_m g) = \lim_{m \to \infty} A_1(f_m) = A_1(1) = 0.$$

Conversely, if $A(g) = 0$ we put $c := \int g \, d\mu$. By Fubini's theorem $\nu_{1/n} = c^{-1/n} g \cdot \mu_{1/n}$ is a nth root of ν_1. The generating functional A_1 induced by ν_1 is

$$A_1(f) = \lim_{n \to \infty} (n(\exp(-\tfrac{1}{n} \log c) \int f \, g \, d\mu_{1/n} - f(0)))$$

$$= A(f\,g) - f(0) \log c, \quad f \in C_{oo}(\mathbb{R}^n) \cap C^2_{lok}(\mathbb{R}^n).$$

Now, $\log c = 0$ follows by considering the functions f_m.

(c) For every $t > 0$, $t\,A(g) = 0$ and $\int g\, d\mu_t = 1$ holds. Hence, $(g \cdot \mu_t)_{t>0}$ is a convolution semigroup. Moreover, this semigroup is vaguely continuous. Since all measures are normed it is also weakly continuous. □

It should be noted that (8.7) (a) can be deduced from Siebert, 1982, who has generalized Kruglov's results.

In the next section infinitely divisible measures on a locally compact semigroup X (e.g. $X = ([-\infty, \infty), +)$) are needed. It is obvious how to adapt Definition (8.1) to this more general situation.

9. The Lévy-Khintchine Formula for Standard Measures

In this section we study infinitely divisible experiments $E = (X, A, (P_\theta)_{\theta \in \Theta})$ for a finite set Θ which is always assumed to be $\Theta = \{1,\ldots,k\}$, $k \geq 2$. We then write $S_k := S_\Theta$. Our first aim is to classify the distributions of the log-likelihood processes of infinitely divisible experiments. We show that E is infinitely divisible if and only if each distribution of the log-likelihood process is infinitely divisible on the semigroup $([-\infty, \infty)^{k-1}, +)$. The different processes are not easy to compare on $[-\infty, \infty)^{k-1}$. Therefore we first consider the P_ℓ-distributions ($\ell = 1,\ldots,k$) of the statistics

$$T = (T_i)_{i=1,\ldots,k}, \quad T_i = \frac{dP_i}{d \sum_{j=1}^{k} P_j}$$

which leads to the theory of standard measures on the simplex S_k. The standard measures of infinitely divisible experiments are described.

<u>(9.1) Composition on simplices.</u> For each $i = 1,\ldots,k$ the set $T_i^k := \{x \in S_k : p_i(x) > 0\}$ is a locally compact Abelian topological semigroup w.r.t. the induced topology, if we put $h_i : (T_i^k)^2 \longrightarrow T_i^k$,

$((x)_r, (y_r)_r) \longmapsto \left(\frac{x_r y_r}{\sum_{j=1}^{k} x_j y_j}\right)_r$ and $xy := h_i(x,y)$. (The multiplication is jointly continuous.)

(i) T_i^k is isomorphic to $([-\infty, \infty)^{k-1}, +)$ equipped with the product of the natural topology of $[-\infty, \infty)$. A topological semigroup isomorphism is (for $i = 1$) defined by

$$\varphi_k (= \varphi_\Theta) : \quad T_1^k \longrightarrow [-\infty, \infty)^{k-1},$$

$$(x_i)_i \longmapsto \left(\log \frac{x_i}{x_1}\right)_{i \geq 2}.$$

φ_k^{-1} is given by

$$\varphi_k^{-1}(y_1,\ldots,y_{k-1}) = \frac{1}{1 + \sum_{i=1}^{k-1} \exp(y_i)} (1, \exp(y_1),\ldots, \exp(y_{k-1})).$$

(ii) Note that $S_k = \bigcup_{i=1}^{k} T_i^k$; $G_k := \bigcap_{i=1}^{k} T_i^k$ is a locally compact topological group being isomorphic to $\mathbb{R}^{k-1}$ and $\varphi_{k|G_k}$ is a group isomorphism.

Sometimes we write T_i^Θ and G_Θ instead of T_i^k and G_k .

(iii) Suppose that $I \in A(\Theta)$ is finite and $i \in I$. Then the restriction of $\varphi_{I,\Theta}$ on T_i^k defines a continuous semigroup homomorphism between T_i^k and T_i^I . By abuse of language let $M_b(T_i^k)$ denote the set of bounded positive measures on S_k which are carried by T_i^k .

<u>(9.2) Convolution on simplices.</u> Assume that μ and ν belong to the same $M_b(T_i^k)$. Then the convolution $\mu * \nu$ is defined as the image measure $(\mu \otimes \nu)^{h_i}$. Obviously, this definition does not depend on the choice of i and j if μ and ν are concentrated on $T_i^k \cap T_j^k$. $(M_b(T_i^k), *)$ is a metrizable topological semigroup if it carries the weak topology of measure theory. Moreover, for every i $M_b(G_k)$ is a subsemigroup of $M_b(T_i^k)$.

<u>(9.3) Lemma.</u> Suppose that P,Q belong to $M_b(S_k)$ and put $P_i := p_i \cdot P$, $Q_i := p_i \cdot Q$ for $i = 1,\ldots,k$. Then $p_i \cdot \sum_{j=1}^k P_j * Q_j = P_i * Q_i$ for each i .

<u>Proof.</u> Let $A \in \mathcal{B}(T_i^k)$ be a Borel set. Then

$$\sum_{j=1}^k \int 1_A p_i \, dP_j * Q_j = \sum_{j=1}^k \int\int 1_A(xy) p_i(xy) dP_j(x) dQ_j(y)$$

$$= \sum_{j=1}^k \int\int 1_A(xy) p_i(xy) p_j(x) p_j(y) dP(x) dQ(y)$$

$$= \int\int 1_A(xy) p_i(xy) \left(\sum_{j=1}^k p_j(x) p_j(y)\right) dP(x) dQ(y)$$

$$= \int\int 1_A(xy) p_i(x) dP(x) p_i(y) dQ(y) = P_i * Q_i(A)$$

if we take into account the definition of xy . □

The next theorem shows why we have introduced the notations of (9.1) and (9.2).

<u>(9.4) Theorem.</u> Assume that $E = (S_k, \mathcal{B}(S_k), (p_i \cdot \mu)_{i=1,\ldots,k})$ and $F = (S_k, \mathcal{B}(S_k), (p_i \cdot \nu)_{i=1,\ldots,k})$ are standard experiments having standard measures μ,ν on S_k . Then

(i) the product experiment $E \otimes F$ has the standard measure $\sum_{j=1}^{k} (p_j \cdot \mu) * (p_j \cdot \nu)$ and

(ii) $(S_k, \mathcal{B}(S_k), ((p_i \cdot \mu) * (p_i \cdot \nu))_{i=1,\dots,k})$ is the standard experiment of $E \otimes F$.

<u>Proof.</u> Let β denote the measure defined in (i) and put $\mu_i := p_i \cdot \mu$, $\nu_i := p_i \cdot \nu$. By (9.3) β is a standard measure and (ii) defines a standard experiment. A simple calculation shows

$$\frac{d(\mu_i \otimes \nu_i)}{d(\sum_{j=1}^{k} \mu_j \otimes \nu_j)}(x,y) = \frac{p_i(x)p_i(y)}{\sum_{j=1}^{k} p_j(x)p_j(y)} 1_{(T_i^k)^2}(x,y) \qquad \sum_{j=1}^{k} \mu_j \otimes \nu_j \quad \text{a.e.}$$

Letting $f_z(x) := \prod_{i=1}^{k} x_i^{z_i}$ we compute the Hellinger transform

$$H(E \otimes F)(z) = \sum_{n=1}^{k} \int_{(T_n^k)^2} f_z(xy)\, d\, \mu_n \otimes \nu_n(x,y)$$

$$= \int_{S_k} f_z \, d\,(\sum_{n=1}^{k} \mu_n * \nu_n) .$$

The uniqueness of the Hellinger transformation for standard experiments implies the result. □

<u>(9.5) Definition.</u> For every $z = (z_i)_i \in S_k$ let $f_z : S_k \longrightarrow [0,1]$ denote the continuous function

$$f_z : x \longmapsto \prod_{i=1}^{k} x_i^{z_i} \quad \text{(we define } 0^0 := 1\text{)}.$$

If $\nu \in M_b(S_k)$ is a bounded measure on S_k, then $z \longmapsto H_\nu(z) := \int f_z \, d\nu$, $z \in S_k$, is the <u>Hellinger transform</u> of ν.

<u>(9.6) Lemma.</u> Suppose that the assumptions of (9.3) hold. Then

$$H_P(z) H_Q(z) = H_{\sum_{j=1}^{k} P_j * Q_j}(z) .$$

<u>Proof.</u>

$$H_{\sum_{j=1}^{k} P_j * Q_j}(z) = \sum_{j=1}^{k} \int\int f_z(xy)\,dP_j(x)\,dQ_j(y)$$

$$= \sum_{j=1}^{k} \int\int f_z((x_i y_i)_i) \frac{1}{\sum_{n=1}^{k} x_n y_n} x_j y_j \, dP(x)\, dQ(y)$$

$$= H_P(z) H_Q(z).$$

□

The next theorem is well-known for the subset of standard measures in $M_b(S_k)$.

<u>(9.7) Theorem.</u>

(i) (Uniqueness theorem) $\nu \longmapsto H_\nu$ is injective on $M_b(S_k)$.

(ii) (Continuity theorem) Suppose that $\mu_n \in M_b(S_k)$, $n \in \mathbb{N}$.

(a) If $\lim_{n\to\infty} H_{\mu_n}(z) = h(z)$ converges pointwise to some function $h: S_k \longrightarrow \mathbb{R}_+$, then there exists a unique measure $\mu \in M_b(S_k)$ such that $H_\mu = h$ and $\mu_n \longrightarrow \mu$, weakly.

(b) Let $\mu \in M_b(\mathbb{R}^k)$. The following assertions are equivalent:

(I) $\mu_n \longrightarrow \mu$, weakly.

(II) $\lim_{n\to\infty} H_{\mu_n}(z) = H_\mu(z)$ for every $z \in S_k$.

<u>Proof.</u> Statement (i) is well-known for standard measures. For its proof suppose $H_\mu = H_\nu$ on S_k.

1. First assume $a_i := \int p_i \, d\mu > 0$ for each i. Then $a_i = \int p_i \, d\nu$ follows by considering $z = (\delta_{ij})_j$. Now, define

$$\mu_i := p_i \cdot \mu, \quad \nu_i := p_i \cdot \nu,$$

$$b := \left(\frac{a_i^{-1}}{\sum_{j=1}^{k} a_j^{-1}} \right)_i, \quad \rho := \sum_{j=1}^{k} a_j^{-1} \varepsilon_b,$$

$$P := \sum_{i=1}^{k} \mu_i * (p_i \cdot \rho) \qquad \text{and}$$

$$Q := \sum_{i=1}^{k} \nu_i * (p_i \cdot \rho).$$

By (9.6) $H_P = H_\mu \cdot H_\rho$, $H_Q = H_\nu \cdot H_\rho$, and (9.3) shows

$$\int p_i \, dP = \int d(\mu_i * (p_i \cdot \rho)) = 1 .$$

Hence, P and Q are standard measures which must coincide since $H_P = H_Q$. Applying (9.3) we finally note that

$$p_i \cdot P = \mu_i * (p_i \cdot \rho) , \quad p_i \cdot Q = \nu_i * (p_i \cdot \rho) ,$$

and therefore $\mu_i = \nu_i$ for every i .

2. We put $J := \{i: \int p_i \, d\mu = 0\}$. If μ does not vanish, then $I = \Theta \setminus J$ is non-void. Since $p_i > 0$ on T_i^k , we conclude:

$$\mu (\bigcup_{j \in J} T_i^k) = \nu (\bigcup_{j \in J} T_i^k) = 0 ,$$

μ and ν are concentrated on $A := \{x \in S_k : x_j = 0 \text{ for all } j \in J\}$. Hence, $\varphi_{I,\Theta}|_A : A \longrightarrow S_I$ is a bijection. Thus, 1. can be applied to the image measures of μ and ν on S_I , and $\mu = \nu$ follows.

Now, we turn to the proof of (ii) (a). Considering $z = (\delta_{ij})_j$ the assumptions yield that $\|\mu_n\| = \sum_{i=1}^{k} \int p_i \, d\mu_n$ is bounded by some $K \in \mathbb{R}_+$. Moreover, we note that $M_b(S_k) \cap \{\nu : \|\nu\| \leq K\}$ is compact and metrizable with respect to the weak topology. Therefore each subsequence of $(\mu_n)_n$ has an accumulation point in $M_b(S_k)$. But every accumulation point μ has the Hellinger transform h . By (i) μ is unique. Thus, $\mu_n \longrightarrow \mu$, weakly. □

A direct proof without using Hellinger transforms of standard measures is contained in Janssen, 1982.

Now, we are able to deal with infinitely divisible experiments. Let E be an experiment for Θ and μ its standard measure.

<u>(9.8) Theorem.</u>

(i) E is infinitely divisible iff each probability measure $\mu_i := p_i \cdot \mu$ is infinitely divisible on T_i^k for $i = 1,\ldots,k$.

(ii) Every infinitely divisible experiment E is embeddable into a continuous semigroup of experiments $(E_t)_{t>0}$ for Θ such that

$$E = E_1 \quad \text{and} \quad H(E_t) = \exp(t \cdot \log H(E)) , \quad t > 0 .$$

The semigroup $(\dot{E}_t)$ is uniquely determined.

(iii) Let (E_t) be a family of experiments having standard measures (μ_t) . The following assertions are equivalent:

(1) $(E_t)_{t>0}$ is a continuous semigroup.

(2) For each $i = 1,\dots,k$ $(p_i \cdot \mu_t)_{t>0}$ is a continuous semigroup of probability measures on T_i^k.

(iv) The set $EI(\Theta)$ of classes of infinitely divisible experiments is weakly compact.

Proof. Statement (i) results from (9.3) and (9.4) and the uniqueness of standard measures.

ad (iv). Suppose that a sequence $(\dot{E}_n)_n$ in $EI(\Theta)$ converges to some $\dot{E} \in E(\Theta)$. Then the Continuity theorem (9.7) implies convergence of the mth roots of $(\dot{E}_n)$ for every fixed $m \in \mathbb{N}$.

ad (ii). Clearly, the statement holds for $t \in \mathbb{Q} \cap (0,\infty)$. Now, (9.7) finishes the proof. □

We note that (9.8) (ii), (iv) are known (cf. LeCam, 1974). Results (9.4) and (9.8) have an interpretation in terms of the corresponding log-likelihood processes (l.p.). For each basis j the distribution of the l.p. of the product experiment $E \otimes F$ is the convolution product of the single l.p.-distributions on $([-\infty,\infty)^{k-1}, +)$. E is infinitely divisible iff each distribution of the l.p. is infinitely divisible on $([-\infty,\infty)^{k-1}, +)$.

Instead of the concept of pairwise imperfectness employed in the first part of this volume we now use the concept of regularity of experiments due to Torgersen (1977, p. 60). As will turn out later (Lemma 9.18) both notions coincide for infinitely divisible experiments.

(9.9) Definition. A finite experiment $E = (X, A, (P_i)_{i=1,\dots,k})$ is regular if there exists a positive measure $\nu \neq 0$ such that $\nu \leq P_i$ for every i. An arbitrary experiment is regular if every restriction to a finite parameter set is regular.

Torgersen remarks: "The statistical interpretation of regularity is, essentially, that one cannot be sure that our observations will exclude some part of Θ". We note that E is regular iff $H(E)(z)$ does not vanish for all $z \in S_k$. To see this assume that $H(E) > 0$ and E is dominated by P. Then it is not hard to check that the measure ν defined

by $\frac{d\nu}{dP} = \min_{1 \le i \le k} \frac{dP_i}{dP}$ satisfies the requirements of Definition (9.9).
The reverse is trivial.

Since all interesting properties of experiments so far depend only on the underlying equivalence classes we assume for the rest of this section that E $((E_t))$ is a standard experiment on S_k (a continuous semigroup of standard experiments). If E_1 is a regular infinitely divisible experiment then $H(E_t)(z)$ converges to 1 for $t \longrightarrow 0$. Hence E_t converges weakly to the totally uninformative experiment E_0 which has the standard measure $k\,\varepsilon_{e_k}$, $e_k := (\frac{1}{k},\dots,\frac{1}{k})$. Consequently, the corresponding convolution semigroups $(p_i \cdot \mu_t)_{t>0}$ converge to ε_{e_k}. A straightforward calculation gives the next statements for Poisson experiments (use (9.3), (9.4)).

<u>(9.10) Example.</u> (Compound Poisson experiments) Let η be a bounded measure on $S_k \setminus \{e_k\}$. For $M_i := p_i \cdot \eta$ we define $M_i^{*0} := \varepsilon_{e_k}$ and

$$E(M_i) := \exp(-\|M_i\|) \sum_{n=0}^{\infty} \frac{M_i^{*n}}{n!} ,$$

the compound Poisson measure of M_i on T_i^k. Put

$$b_t := \left(\left(\frac{\exp(-t\,\|M_i\|)}{\sum_{j=1}^{k} \exp(-t\,\|M_j\|)} \right)_i \right) \in G_k .$$

(i) For every i $\mu_{i,t} := \varepsilon_{b_t} * E(t\,M_i)$, $t \ge 0$, is a continuous convolution semigroup on T_i^k.

(ii) $E_t = (S_k, \mathcal{B}(S_k), (\mu_{i,t})_{i=1,\dots,k})$, $t \ge 0$, is a continuous semigroup of regular standard experiments having Hellinger transforms
$H(E_t)(z) = \exp(t \int \psi_z \, d\eta)$ where $\psi_z(x) := f_z(x) - \sum_{i=1}^{k} z_i\, p_i(x)$, $x \in S_k$.
We call $(E_t)_t$ a <u>compound Poisson semigroup</u> (resp. E_t a <u>compound Poisson experiment</u>) and η the <u>Lévy measure</u> of the semigroup.

<u>(9.11) Theorem.</u> Let E be a regular infinitely divisible standard experiment. Then E can be decomposed as a product of two regular standard experiments E_1 and E_2 such that $E \sim E_1 \otimes E_2$, E_1, E_2 are unique and

(i) E_1 is homogeneous and infinitely divisible,

(ii) E_2 is a compound Poisson experiment whose Lévy measure η_2 satisfies $\eta_2(G_k) = 0$.

<u>Proof.</u> Let (μ_t) denote the standard measures of the continuous semigroup induced by E (cf. (9.8)). We put $\mu_{i,t} := p_i \cdot \mu_t$. Since E is regular we obtain $\mu_t(G_k) > 0$ and $\mu_{i,t}(G_k) > 0$ because of $H_{\mu_t}(e_k) > 0$. We introduce

$$\rho_{i,t} := \mu_{i,t|G_k} , \qquad \nu_{i,t} := \mu_{i,t|CG_k}$$

$$\rho_t := \sum_{j=1}^{k} \rho_{i,t} \quad \text{and} \quad \nu_t := \sum_{j=1}^{k} \nu_{i,t}.$$

Since $\mu_{i,t} = p_i \cdot \mu_t$ we conclude

$$p_i \cdot \rho_t = \rho_{i,t} \quad \text{and} \quad p_i \cdot \nu_t = \nu_{i,t} \; .$$

Applying (9.8) (iii) and $M_b(T_i^k \setminus G_k) * M_b(T_i^k) \subset M_b(T_i^k \setminus G_k)$ we obtain $\rho_{i,t} * \rho_{i,s} = \rho_{i,t+s}$ for all $s,t > 0$. Hence, $\|\rho_{i,t}\| = \exp(-ta_i)$ for some $a_i \geq 0$. By (9.4) the Hellinger transforms satisfy the equation $H_{\rho_{t+s}}(z) = H_{\rho_t}(z) H_{\rho_s}(z)$. Therefore there exists a function $g: S_k \longrightarrow (-\infty, 0]$ such that $H_{\rho_t}(z) = \exp(tg(z))$.
Now, $H_{\mu_t}(z) = H_{\rho_t}(z) + H_{\nu_t}(z)$ proves

$$\lim_{t\to 0} \frac{\exp(t \log H(E)(z)) - 1}{t} = \lim_{t\to 0} \frac{\exp(tg(z)) - 1}{t} + \lim_{t\to 0} \frac{H_{\nu_t}(z)}{t} .$$

Consequently, the Continuity theorem shows that $\frac{\nu_t}{t}$ converges weakly to some measure η_2 concentrated on $S_k \setminus G_k$. Because of $\|\nu_t\| = \sum_{j=1}^{k} (1-\exp(-ta_j))$, $\|\eta_2\| = \sum_{j=1}^{k} a_j$. Put

$$b_t := \left(\left(\frac{\exp(ta_i)}{\sum_{j=1}^{k} \exp(ta_j)} \right)_i \right), \quad t \geq 0 .$$

In the following we shall use (9.3) and (9.4) several times. First we remark that $\beta_t = \sum_{j=1}^{k} \exp(ta_j)\, \varepsilon_{b_t} * \rho_{i,t}$ is a standard measure. This can be proved by taking into consideration $P_t = \sum_{j=1}^{k} \exp(ta_j)\, \varepsilon_{b_t}$ and $Q_t = \rho_t$ because of $p_i \cdot \beta_t = \exp(ta_i)\, \varepsilon_{b_t} * \rho_{i,t}$.

Put $(E_1)_t := (S_k, \mathcal{B}(S_k), (p_i \cdot \beta_t)_{i=1,\dots,k})$ and let E_2 denote the compound Poisson experiment with Hellinger transform

$$H(E_2)(z) = \exp(\int \psi_z \, d\eta_2) .$$

E_1 is infinitely divisible and homogeneous since β_1 is concentrated on G_k .

Now, $H(E_1)(z) = H_{P_1}(z) H_{\rho_1}(z)$ and $H_{P_1}(z) = \sum_{i=1}^{k} \exp(a_i z_i)$. Hence,

$$H(E_1)(z) \cdot H(E_2)(z) = H_{\rho_1}(z) \exp(\sum_{i=1}^{k} a_i z_i + \int \psi_z \, d\eta_2)$$

$$= \exp(g(z) + H_{\eta_2}(z)) = H(E)(z) ,$$

if we take into account $g(z) + H_{\eta_2}(z) = \log H(E)(z)$.

Uniqueness of the decomposition: Suppose $\mu_{i,t}$ is the convolution product of an infinitely divisible measure on G_k (arising from E_1) and $E(tp_i \cdot \eta_2') * \varepsilon_{b_t}$ such that $\eta_2'(CG_k) = 0$. Then it is easy to see that $\frac{1}{t} \mu_{i,t|CG_k}$ converges weakly to $p_i \cdot \eta_2'$. Hence, $\eta_2 = \eta_2'$. □

<u>(9.12) Example.</u> (Homogeneous experiments) There is a one-to-one correspondence between the set of classes of infinitely divisible, homogeneous experiments

$$W := \{\dot{E} \in EI(\Theta): E \text{ homogeneous} \}$$

and the following set of infinitely divisible measures on $\mathbb{R}^{k-1}$

$$Z := \{\rho \in M_1(\mathbb{R}^{k-1}): \rho \text{ infinitely divisible, } \int \exp(y_i) d\rho(y) = 1 \text{ for each } i = 1,\dots, k-1 \} .$$

Suppose that μ is a standard measure of $\dot{E} \in W$. Then the correspondence $\mu \longleftrightarrow (p_1 \cdot \mu)^{\varphi_k} = \rho$ satisfies the assertion above. Note that for $i \geq 2$

$$1 = \int p_i \, d\mu = \int \frac{p_i}{p_1} \circ \varphi_k^{-1} \, d(p_1 \cdot \mu)^{\varphi_k}$$

and

$$\exp(y_{i-1}) = \frac{p_i}{p_1} \circ \varphi_k^{-1} (y_1,\dots, y_{k-1})$$

implies $\rho \in Z$.

Conversely, we remark that by (8.7) all measures $\rho_1 = \rho \in Z$ and ρ_i are infinitely divisible if we define ρ_i by the ρ-density $y \longmapsto \exp(y_{i-1})$ on $\mathbb{R}^{k-1}$, $i \geq 2$.

Since $\left(\left(\frac{d\rho_i}{d \sum_{j=1}^{k} \rho_j} (y) \right)_i \right) = \varphi_k^{-1}(y)$ the standard representation of

$F = (\mathbb{R}^{k-1}, \mathcal{B}(\mathbb{R}^{k-1}), (\rho_i)_{i=1,\dots,k})$ consists of the measures $(\rho_i)^{\varphi_k^{-1}}$ on S_k. By (9.8) the class $\dot{F}$ belongs to W.

Observe that every $\dot{E} \in W$ can be described by one single infinitely divisible measure ρ defined on $\mathbb{R}^{k-1}$. Let $E = (X, A, (P_i)_{i=1,\dots,k})$ be an arbitrary homogeneous experiment having a standard measure μ. Then the experiment F constructed above remains equivalent to E (this does not depend on the infinite divisibility of E). Moreover, ρ_i is the distribution of the l.p. $(\log \frac{dP_j}{dP_1})_{j \neq 1}$ w.r.t. P_i.

Now, we change the basis of the l.p. and consider for $i \neq 1$ the vector

$$v = (\log \frac{dP_1}{dP_i}, \dots, \log \frac{dP_{i-1}}{dP_i}, \log \frac{dP_{i+1}}{dP_i}, \dots, \log \frac{dP_k}{dP_i}) \in \mathbb{R}^{k-1}$$

w.r.t. P_i. Then there is a linear transformation A on $\mathbb{R}^{k-1}$ which maps the l.p. w.r.t. basis 1 to the vector

$$v = (\log \frac{dP_j}{dP_i} = \log \frac{dP_j}{dP_1} - \log \frac{dP_i}{dP_1})_{j \neq i}.$$

The distribution of v under P_i equals the image distribution ρ_i^A.

Finally, we note that A preserves infinite divisibility and normality of distributions. If E is infinitely divisible (or a Gaussian experiment) each log-likelihood distribution is infinitely divisible (normal, respectively).

(9.13) Notation. Let us introduce the following notations.

$$C^2_{lok}(T_i^k) := \{f \in C_b(T_i^k) : f \circ \varphi^{-1}_{k|\mathbb{R}^{k-1}} \in C^2_{lok}(\mathbb{R}^{k-1})\},$$

$$C^2_{lok}(S_k) := \{f \in C_b(S_k) : f \circ \varphi^{-1}_{k|\mathbb{R}^{k-1}} \in C^2_{lok}(\mathbb{R}^{k-1})\},$$

$$D_i f := \frac{\partial}{\partial y_i} f \circ \varphi_k^{-1}(0), \quad D_i D_j f := \frac{\partial}{\partial y_i} \frac{\partial}{\partial y_j} f \circ \varphi_k^{-1}(0)$$

for suitable f, $1 \le i \le k-1$. We define $g_i, g: \mathbb{R}^{k-1} \longrightarrow \mathbb{R}$ by $g_i(y) := \exp(y_i)$ and $g(y) := 1 + \sum_{i=1}^{k-1} g_i(y)$.

(9.14) Theorem. Let E be infinitely divisible and let $(E_t)_t$ be the corresponding continuous semigroup of experiments having standard measures (μ_t). If E is regular the following assertions are valid for $\mu_t^i := p_i \cdot \mu_t$:

(i) $B_i(f) := \lim_{t\to 0} \dfrac{\int f\, d\mu_t^i - f(e_k)}{t}$ exists for every $f \in C^2_{lok}(T_i^k)$.

(ii) Let $E \sim E_1 \otimes E_2$ denote the decomposition according to (9.11). Suppose that ν is the standard measure of E_1 and the semigroup induced by $(p_1 \cdot \nu)^{\varphi_k}$ on $\mathbb{R}^{k-1}$ has the generating functional A (with the Lévy-Khintchine representation $(b', (a_{ij}), \eta_1)$).

Put $\tilde{\eta} := \eta_1^{\varphi_k^{-1}} + p_1 \cdot \eta_2$. Then B_1 admits the following representation on $C^2_{lok}(T_1^k)$:

$$B_1(f) = \langle (D_i f)_i, b \rangle + \sum_{i,j=1}^{k} a_{ij} D_i D_j f$$

$$+ \int_{S_k \setminus \{e_k\}} \left(f(x) - f(e_k) - \sum_{i=1}^{k-1} D_i f \left(\frac{x_{i+1}}{x_1} - 1\right)\right) d\tilde{\eta}(x).$$

Moreover: $b = (- a_{ii} - ((p_{i+1} \cdot \eta_2)(\{p_1 = 0\})))_{i=1,\dots,k-1}$.

<u>Proof.</u> Let (ν_t) be the family of standard measures belonging to the semigroup induced by E_1. Put $\nu_t^i := p_i \cdot \nu_t$ and $a_i := \int p_i \, d\eta_2$. Then by (9.3) and (9.10):

$$\mu_t^i = \nu_t^i * \varepsilon_{b_t} * E(tp_i \cdot \eta_2) \quad \text{and} \quad b_t = \left(\left(\frac{\exp(-ta_i)}{\sum_{j=1}^{k} \exp(-ta_j)} \right)_i \right).$$

For $i = 1$ we have

$$\mu_t^1 = \exp(-ta_1) \left[\nu_t^1 * \varepsilon_{b_t} + t(\nu_t^1 * \varepsilon_{b_t}) * p_1 \cdot \eta_2 + \dots \right]$$

and

$$(\nu_t^1 * \varepsilon_{b_t})^{\varphi_k} = (\nu_t^1)^{\varphi_k} * \varepsilon_{t(a_1 - a_2, a_1 - a_3, \dots, a_1 - a_k)}.$$

Hence,

$$\frac{\int f\, d\mu_t^1 - f(e_k)}{t} = \exp(-ta_1) \frac{\int f \circ \varphi_k^{-1}\, d(\nu_t^1 * \varepsilon_{b_t})^{\varphi_k} - f \circ \varphi_k^{-1}(0)}{t}$$

$$+ f(e_k) \frac{\exp(-ta_1) - 1}{t} + \exp(-ta_1) \int f\, d(\nu_t^1 * \varepsilon_{b_t} * p_1 \cdot \eta_2) + o(t).$$

For each term of the sum the $\lim_{t \to 0}$ exists. Therefore, $B_1(f)$ exists and has the form

$$B_1(f) = A(f \circ \varphi_k^{-1}) + \langle (D_i f)_i, (a_1-a_2, a_1-a_3, \ldots, a_1-a_k) \rangle$$
$$- f(e_k)\, a_1 + \int f\, d(p_1 \cdot \eta_2)$$

because the first two terms are the generating functional of $\nu_t^1 * \varepsilon_{b_t}$. Observe that $\nu_t^1 * \varepsilon_{b_t} \longrightarrow \varepsilon_{e_k}$ weakly. Inserting the definition of a_i we obtain

$$a_1 - a_i = \int (x_1 - x_i)\, d\eta_2(x) = \int \left(1 - \frac{x_i}{x_1}\right) d\,(p_1 \cdot \eta_2)(x) - (p_i \cdot \eta_2)(\{p_1 = 0\})$$

$$B_1(f) = A(f \circ \varphi_k^{-1}) + \int_{S_k \setminus \{e_k\}} \left(f(x) - f(e_k) - \sum_{i=1}^{k-1} D_i f \left(\frac{x_{i+1}}{x_1} - 1\right)\right) d(p_1 \cdot \eta_2)(x)$$
$$- \langle (D_i f)_i, ((p_{i+1} \cdot \eta_2)(\{p_1 = 0\}))_i \rangle \quad .$$

In view of (8.7) $A(g_i)$ exists for each i.

Hence, $y \longmapsto g_i(y) - 1 - \dfrac{y_i}{1 + \|y\|^2}$ is η_1-integrable on $\mathbb{R}^{k-1}$.

Therefore, supposing that f belongs to $C^2_{lok}(T_i^k)$, we can write the integral term of the canonical representation of A in the following form:

$$A(f \circ \varphi_k^{-1}|_{\mathbb{R}^{k-1}}) = \langle (D_i f)_i, b' \rangle + \sum_{i,j=1}^{k-1} a_{ij} D_{ij} f$$
$$+ \int_{\mathbb{R}^{k-1} \setminus \{0\}} \left(f \circ \varphi_k^{-1}(y) - f(e_k) - \frac{\langle (D_i f)_i, y \rangle}{1 + \|y\|^2}\right) d\eta_1(y)$$
$$= \langle (D_i f)_i, b'' \rangle + \sum_{i,j=1}^{k-1} a_{ij} D_{ij} f$$
$$+ \int_{\mathbb{R}^{k-1} \setminus \{0\}} \left(f \circ \varphi_k^{-1}(y) - f(e_k) - \sum_{i=1}^{k-1} D_i f(g_i(y) - 1)\right) d\eta_1(y)$$

for a suitable $b'' = (b_i'')_i \in \mathbb{R}^{k-1}$. For the last term the transformation theorem yields

$$\int_{S_k \setminus \{e_k\}} \left(f(x) - f(e_k) - \sum_{i=1}^{k-1} D_i f \left(\frac{x_{i+1}}{x_1} - 1\right)\right) d\eta_1^{\varphi_k^{-1}}(x) \; .$$

If we substitute g_i for $f \circ \varphi_k^{-1}|_{\mathbb{R}^{k-1}}$, then Lemma (8.7) implies $0 = A(g_i) = b_i'' + a_{ii}$. Similar arguments prove that B_i exists for arbitrary i. □

Our theorem shows that

$$\lim_{t\to 0} \frac{\int f\, d\mu_t - kf(e_k)}{t} = \sum_{i=1}^{k} B_i(f) =: B(f)$$

exists for $f \in C^2_{lok}(S_k)$. We call B the generating functional of the semigroup E_t. In the following discussion we want to give a representation of the "derivative of the standard measure μ_t" at $t = 0$. For this it is useful to introduce a new function space which utilizes the analytic structure of $\mathbb{R}_+^k \supset S_k$. For example the important functions f_z can be extended on $\mathbb{R}_+^k$. We consider functions h of the subsequent type:

Notation. Let $U \subset \mathbb{R}^k$ be an open neighbourhood of e_k and $h: S_k \cup U \longrightarrow \mathbb{R}^1$ a function which is twice continuously differentiable on U and continuous on S_k. Let us denote by $C^2_{lok}(S_k, \mathbb{R}^k)$ the set of functions h for which such a neighbourhood U exists; plainly, $h|_{S_k} \in C^2_{lok}(S_k)$. Conversely, if $f \in C^2_{lok}(S_k)$ is such that $f \circ \varphi_k^{-1}$ is twice continuously differentiable on some open neighbourhood $V \subset \mathbb{R}^{k-1}$ of e_k, then we construct a function $h \in C^2_{lok}(S_k, \mathbb{R}^k)$ whose restriction on S_k coincides with f. Put $\varphi: (0,\infty)^k \longrightarrow \mathbb{R}^{k-1}$, $\varphi(x) := (\log \frac{x_i}{x_1})$ $i \geq 2$. Then $U = \varphi^{-1}(V)$ is open; let $h(x) := (f \circ \varphi_k^{-1})(\varphi(x))$ if $x \in U$ and $h(x) := f(x)$ if $x \in S_k \setminus U$. We note that this extension of f is not unique.

Now, we are able to use the partial derivatives h_{x_i} and $h_{x_i x_j}$ in $\mathbb{R}^k$.

(9.15) Theorem. (Lévy-Khintchine theorem for standard measures) Assume that $(E_t)_{t>0}$ is a continuous semigroup of regular experiments having standard measures (μ_t). Then:

(a) For every $h \in C^2_{lok}(S_k, \mathbb{R}^k)$ the limit

$$B(h) = \lim_{t\to 0} \frac{\int h\, d\mu_t - k\,h(e_k)}{t}$$

exists.

We call B the generating functional of $(E_t)_t$. B has the canonical form

$$B(h) = \langle \nabla(g(h \circ \varphi_k^{-1}))(0), b \rangle + \sum_{i,j=1}^{k-1} a_{ij} \frac{\partial^2}{\partial y_1 \, \partial y_j} (g(h \circ \varphi_k^{-1}))(0)$$

$$+ \int_{S_k \setminus \{e_k\}} (h(x) - h(e_k) - \sum_{i=1}^{k} (x_i - \tfrac{1}{k}) h_{x_i}(e_k)) \, d\eta(x) .$$

(For the definition of g compare (9.13).)

(i) $(a_{ij})_{i,j=1,\dots,k-1}$ is a real positive semi-definite matrix,

(ii) $b = (-a_{ii})_{i=1,\dots,k-1} \in \mathbb{R}^{k-1}$,

(iii) η is a Radon measure on $S_k \setminus \{e_k\}$ defined by

$$\int h \, d\mu = \lim_{t \to 0} \frac{1}{t} \int h \, d\mu_t \quad \text{for every} \quad h \in C_{oo}(S_k \setminus \{e_k\})$$

and η satisfies

(iv) $$\int_{S_k \setminus \{e_k\}} \sum_{i=1}^{k} (x_i - \tfrac{1}{k})^2 \, d\eta(x) < \infty .$$

(b) Conversely, if a triplet $(b, (a_{ij})_{i,j}, \eta)$ satisfies (i), (ii) and (iv), then B defines the generating functional of a uniquely determined semigroup of standard experiments E_t on S_k having the representation B.

<u>(9.16) Definition.</u> We call η the <u>Lévy measure</u> of the semigroup (of experiments). In general, every measure η on S_k satisfying (9.15) (iv) and $\eta\{e_k\} = 0$ is a <u>Lévy measure</u> on S_k.

<u>Proof of Theorem (9.15).</u> First we demonstrate the following two statements.

1) Let η be a Radon measure on $\mathbb{R}^{k-1} \setminus \{0\}$. Then η is a Lévy measure in the sense of (8.3) iff $\eta^{\varphi_k^{-1}}$ satisfies (iv).

2) Let η be a Radon measure on $S_k \setminus G_k$. Then $\eta(S_k) < \infty$ iff η satisfies (iv).

1. If we define $W := \{y \in \mathbb{R}^{k-1} : \|y\| < 1\}$, then there is a $K > 0$ such that

$$K \, 1_{\varphi_k^{-1}(CW)}(x) \leq \sum_{i=1}^{k} (x_i - \tfrac{1}{k})^2 \leq k .$$

Hence, $\eta(CW) < \infty$ iff $\int_{\varphi_k^{-1}(CW)} \sum_{i=1}^{k} (x_i - \frac{1}{k})^2 \, d\eta^{\varphi_k^{-1}}(x) < \infty$.

Moreover,

$$p_1(\varphi_k^{-1}(y)) \geq ((k-1)e+1)^{-1} =: a, \quad y \in W,$$

and

$$\int_W \|y\|^2 \, d\eta(y) < \infty \quad \text{iff} \quad \int_W \sum_{i=1}^{k-1} (\exp(y_i) - 1)^2 \, d\eta(y) < \infty,$$

since $(\exp(y_i) - 1)^2$ behaves like y_i^2 near 0. Now,

$$\int_W \sum_{i=1}^{k-1} (\exp(y_i) - 1)^2 \, d\eta(y) = \int_{\varphi_k^{-1}(W)} \frac{1}{x_1^2} \sum_{i=2}^{k} (x_i - x_1)^2 \, d\eta^{\varphi_k^{-1}}(x).$$

As $1 \leq \frac{1}{x_1^2} \leq \frac{1}{a^2}$ for $x \in \varphi_k^{-1}(W)$, $\eta|_W$ is a Lévy measure iff

$$\int_{\varphi_k^{-1}(W)} \sum_{i=2}^{k} (x_i - x_1)^2 \, d\eta^{\varphi_k^{-1}}(x) < \infty.$$

Finally, we remark that

$$\sum_{i=2}^{k} (x_i - x_1)^2 = \sum_{i=2}^{k} (x_i - \frac{1}{k})^2 + (k+1)(x_1 - \frac{1}{k})^2$$

on S_k.

2. The second assertion follows from

$$(\frac{1}{k})^2 \leq \sum_{i=1}^{k} (x_i - \frac{1}{k})^2 \leq k, \quad x \in S_k \setminus G_k.$$

3. First we assume that E_t is homogeneous. Then $((p_1 \cdot \mu_t)^{\varphi_k})_t$ is a continuous convolution semigroup on $\mathbb{R}^{k-1}$ having a generating functional A. We note that $(p_i \cdot \mu_t)^{\varphi_k} = g_{i-1} \cdot (p_1 \cdot \mu_t)^{\varphi_k}$ for $i \geq 2$, (cf. (9.12)). Then (8.7) and the calculation in the proof of (9.14) imply

$$B(h) = \sum_{i=1}^{k} \lim_{t \to 0} \frac{\int h \, d\mu_t^i - h(e_k)}{t} = A(h \circ \varphi_k^{-1}) + \sum_{i=1}^{k-1} A(g_i (h \circ \varphi_k^{-1}))$$

$$= A(g(h \circ \varphi_k^{-1})).$$

By (8.7) the function gf lies in $C^2_{lok,A}(\mathbb{R}^{k-1})$ for every $f \in C^2_{lok}(\mathbb{R}^{k-1})$

and $g1_{\{y:\, \|y\| > 1\}}$ is η_1-integrable where η_1 denotes the Lévy measure of the integral term of $\dot{A}$. According to the proof of (9.14) A has the representation

$$(*) \qquad A(f) = \langle \nabla f(0), b \rangle + \sum_{i,j=1}^{k-1} a_{ij} \frac{\partial^2}{\partial y_i\, \partial y_j} f(0)$$
$$+ \int_{\mathbb{R}^{k-1} \setminus \{0\}} (f - f(0) - \langle \nabla f(0), (g_i - 1)_i \rangle)\, d\eta_1 ,$$
$$f \in C^2_{lok,A}(\mathbb{R}^{k-1}) .$$

Putting $f := h \circ \varphi_k^{-1}\big|_{\mathbb{R}^{k-1}}$, the canonical representation of the first and second term of B is obvious, and (i), (ii) are true.

Now, we calculate the integral part. We observe that $\tilde{\eta} := g \cdot \eta_1$ is a Lévy measure on $\mathbb{R}^{k-1}$.

$$\int_{\mathbb{R}^{k-1} \setminus \{0\}} (gf - kf(0) - \langle \nabla(gf)(0), (g_i - 1)_i \rangle)\, d\eta_1 =$$

$$\int_{G_k} \left(h(x) - \frac{1}{1 + \sum_{i=2}^{k} \frac{x_i}{x_1}} \left(kh(e_k) + \sum_{i=1}^{k-1} \frac{\partial}{\partial y_1} (gf)(0) \left(\frac{x_{i+1}}{x_1} - 1\right)\right)\right) d\tilde{\eta}^{\varphi_k^{-1}}(x) .$$

Let $H(x)$ denote the last integrand above. We note that φ_k^{-1} has the following Jacobian at 0 (we get a $k\times(k-1)$ matrix)

$$-\frac{1}{k^2} \begin{pmatrix} 1 & \cdots & 1 \\ 1 & & 1 \\ & 1 & \\ \vdots & \ddots & \vdots \\ 1 & \cdots & 1 \end{pmatrix} + \frac{1}{k} \begin{pmatrix} 0 & \cdots & 0 \\ 1 & & 0 \\ & 1 & \\ \vdots & \ddots & \vdots \\ 0 & \cdots & 1 \end{pmatrix}$$

and

$$\nabla(gf) = h(e_k)\,(1,\ldots,1) + (-k)\nabla f(0)$$
$$= \left(h(e_k) - \frac{1}{k} \sum_{i=1}^{k} h_{x_i}(e_k)\right)(1,\ldots,1) + (h_{x_2}(e_k),\ldots,h_{x_k}(e_k))$$

where $(1,\ldots,1) \in \mathbb{R}^{k-1}$. Hence, $H(x)$ has the form

$$H(x) = h(x) - kh(e_k)x_1 - \sum_{i=1}^{k-1} \frac{\partial}{\partial y_i} (gf)(0)(x_{i+1} - x_1)$$

$$= h(x) - h(e_k)\left(kx_1 - (k-1)x_1 + \sum_{i=1}^{k-1} x_{i+1}\right)$$

$$+ \left(\frac{1}{k} \sum_{i=1}^{k} h_{x_i}(e_k)\right) \sum_{i=2}^{k} (x_i - x_1) - \sum_{i=2}^{k} h_{x_i}(e_k)(x_i - x_1) .$$

Now, $\sum_{i=1}^{k-1} x_{i+1} = 1 - x_1$ and $\sum_{i=2}^{k} (x_i - x_1) = 1 - kx_1$, showing that

$$H(x) = h(x) - h(e_k) - \sum_{i=1}^{k} (x_i - \frac{1}{k})\, h_{x_i}(e_k)\ .$$

Then $B(h)$ has the desired form with $\eta := \tilde{\eta}^{\varphi_k^{-1}}$. Moreover, (iii) and (iv) and the uniqueness of A follow.

Conversely, if the triplet in (b) has the properties mentioned above and η is concentrated on G_k, then we first put $\eta_1 := \frac{1}{g} \cdot \eta^{\varphi_k}$. Now, by (*) above and $(b, (a_{ij}), \eta_1)$ we define a generating functional A on $C^2_{lok}(\mathbb{R}^{k-1})$ which gives a continuous convolution semigroup (ρ_t). By (9.12) and the above calculation the standard experiment of $(\mathbb{R}^{k-1}, \mathcal{B}(\mathbb{R}^{k-1}), (\rho_t, g_1 \cdot \rho_t, g_2 \cdot \rho_t, \ldots, g_{k-1} \cdot \rho_t))$ has the generating functional B.

4. The general case. By (9.11) there are a continuous semigroup of homogeneous experiments $E_t^{(1)}$ and a Poisson semigroup $E_t^{(2)}$ with Lévy measure η_2 such that $E_t \sim E_t^{(1)} \otimes E_t^{(2)}$. If we use the notation of (9.14), then $\mu_t^i = \nu_t^i * \varepsilon_{b_t} * E(t\,\eta_2)$.

Let $\tilde{B}$ be the generating functional of $E_t^{(1)}$. Then for each i:

$$\lim_{t\to 0} \frac{\int h\, d\nu_t^i * \varepsilon_{b_t} - h(e_k)}{t} = \lim_{t\to 0} \frac{\int h \circ \varphi_k^{-1}\, d(\nu_t^i * \varepsilon_{b_t})^{\varphi_k} - h(e_k)}{t}$$

$$= \lim_{t\to 0} \frac{\int h \circ \varphi_k^{-1}\, d\nu_t^{i\,\varphi_k} - h(e_k)}{t} + \frac{h \circ \varphi_k^{-1} (\varphi_k(b_t)) - h(e_k)}{t}\ .$$

Now,

$$\lim_{t\to 0} \sum_{i=1}^{k} \frac{\int h\, d\nu_t^i * \varepsilon_{b_t} - h(e_k)}{t} = \tilde{B}(h) + k \lim_{t\to 0} \frac{h(b_t) - h(e_k)}{t}\ .$$

If we consider the decomposition of μ_t (see (9.14))

$$\mu_t = \sum_{i=1}^{k} \exp(-ta_i)\,(\nu_t^i * \varepsilon_{b_t} + t(\nu_t^i * \varepsilon_{b_t}) * p_i \cdot \eta_2 + \ldots),$$

then

$$\frac{\int h\, d\mu_t - kh(e_k)}{t} = \sum_{i=1}^{k} \Bigg[\exp(-ta_i) \frac{\int h\, d\nu_t^i * \varepsilon_{b_t} - h(e_k)}{t}$$

$$+ h(e_k) \frac{\exp(-ta_i) - 1}{t} + \exp(-ta_i) \int h\, d(\nu_t^i * \varepsilon_{b_t} * p_i \cdot \eta_2) + o(t) \Bigg].$$

For $t \longrightarrow 0$ we get the limit

$$\tilde{B}(h) + k \lim_{t\to 0} \frac{h(b_t) - h(e_k)}{t} - \sum_{i=1}^{k} a_i h(e_k) + \sum_{i=1}^{k} \int h p_i \, d\eta_2 .$$

Because of

$$\frac{\partial}{\partial t} \left(\frac{\exp(-ta_i)}{\sum_{j=1}^{k} \exp(-ta_j)} \right) \Bigg|_{t=0} = \frac{1}{k^2} (-a_i k + \sum_{j=1}^{k} a_j)$$

the chain-rule yields the following expression for the second term:

$$k \sum_{i=1}^{k} h_{x_i}(e_k) \left(- \frac{1}{k} a_i + \frac{1}{k^2} \sum_{j=1}^{k} a_j \right) .$$

Taking into consideration $a_i = \int x_i \, d\eta_2(x)$ and $\sum_{i=1}^{k} p_i = 1$, the derivative has the form

$$\tilde{B}(h) + \int \left(h(x) - h(e_k) - \sum_{j=1}^{k} (x_j - \frac{1}{k}) h_{x_j}(e_k) \right) d\eta_2(x) .$$

Together with 3. this proves (a).

Conversely, if the assumptions of (b) are satisfied, then $(b, (a_{ij})_{i,j}, \eta_{|G_k})$ determines a homogeneous semigroup $E_t^{(1)}$. Moreover, by 2) $\eta_{|CG_k}$ is finite and defines a Poisson semigroup $E_t^{(2)}$. Altogether we observe that the class of $E_t = E_t^{(1)} \otimes E_t^{(2)}$ has the desired properties. □

The semigroup is homogeneous iff the Lévy measure is concentrated on G_k.

In 1974 LeCam proved a canonical representation of the Hellinger transforms of infinitely divisible experiments. These results can also be derived from (9.15).

<u>(9.17) Corollary.</u> (Lévy-Khintchine formula for Hellinger transforms)

(a) For every infinitely divisible regular experiment E the Hellinger transform has the form $H(E)(z) = \exp(\Psi(z))$, $z \in S_k$, where

$$\Psi(z) := \sum_{i,j=1}^{k} c_{ij} z_i z_j - \sum_{i=1}^{k} c_{ii} z_i + \int_{S_k \setminus \{e_k\}} \psi_z \, d\eta .$$

(i) $(c_{ij})_{i,j \geq 2}$ is a real positive semi-definite matrix and $c_{i1} = c_{1j} = 0$ for all $i,j \in \Theta$.

(ii) η is a Lévy measure on S_k .

(b) Conversely, if a pair $((c_{ij})_{i,j}, \eta)$ satisfies (i) and (ii), then $\exp(\Psi(z))$ is the Hellinger transform of an infinitely divisible regular experiment E . The class $\dot{E}$ is unique.

Proof. We suppose that E is embedded into (E_t) (cf. Theorem (9.8)). Then we extend the definition of f_z to $\mathbb{R}_+^k$ and note that $f_z \in C^2_{lok}(S_k, \mathbb{R}^k)$, $\Psi(z) = B(f_z)$ (see (9.15)). Now, using (9.15) (i) we define $c_{ij} := 0$ if $j=1$ or $i=1$ and $c_{ij} := a_{i-1,j-1}$. Because of $(f_z)_{x_i}(e_k) = z_i$, $\sum_{i=1}^{k} z_i = 1$ and $g(f_z \circ \varphi_k^{-1})(y) = \exp(\sum_{i=2}^{k} z_i y_{i-1})$ assertion (a) is proved.

ad (b). The triplet $((-c_{i+1,i+1})_{i=1,\dots,k-1}, (c_{i+1,j+1})_{i,j=1,\dots,k-1}, \eta)$ defines a semigroup (E_t) satisfying $H(E_1)(z) = \exp(\Psi(z))$. □

Now, we come back to the concept of regularity of experiments (cf. Definition (9.9)). For infinitely divisible experiments this notion coincides with LeCam's notion of pairwise imperfectness, as we shall show now.

(9.18) Lemma. Let E be infinitely divisible. The following assertions are equivalent:

(i) E is regular.

(ii) There exists a parameter $i \in \Theta$ such that $E_{\{i,j\}}$ is regular for each $j \neq i$.

Proof. (ii) $\Longrightarrow$ (i). Let (E_t) denote the semigroup in which E is embedded. Then $\dot{E}_t$ converges weakly to some $\dot{F} \in \mathcal{E}(\Theta)$ as $t \longrightarrow 0$, and $H(F)(z) = 1$ if $\log H(E)(z)$ is not equal to $-\infty$ and $H(F)(z) = 0$, otherwise. Furthermore, $(E_t)_{\{i,j\}}$ is regular and converges to the trivial experiment. Hence, $F_{\{i,j\}}$ is trivial showing that F is trivial. Therefore, $\Psi > -\infty$ and E is regular. □

The assertion of the next theorem is known, compare with LeCam, 1974, Part 8, Prop. 2. By this theorem it suffices to examine regular subexperiments of infinitely divisible experiments.

(9.19) Theorem. Let $E = (X, A, (P_i)_{i=1,\dots,k})$ be infinitely divisible. Then there exists a unique partition $\Theta_1,\dots,\Theta_r$ of Θ into r non-void parts such that

(i) E_{Θ_j} is regular, $1 \le j \le r$,

(ii) P_t and P_s are mutually orthogonal if t and s lie in different sets Θ_j.

Proof. We define

$$\tilde{\Theta}_i := \{j \in \Theta : P_i \text{ and } P_j \text{ not mutually orthogonal}\}.$$

Then (9.18) implies:

$$\tilde{\Theta}_i = \tilde{\Theta}_j \quad \text{if} \quad j \in \tilde{\Theta}_i$$

and

$$\tilde{\Theta}_i \cap \tilde{\Theta}_j = \emptyset \quad \text{if} \quad j \notin \tilde{\Theta}_i. \qquad \square$$

Now, we deal with arbitrary continuous semigroups (E_t) of experiments. Then the decomposition (9.19) is independent of t and $(E_t)_{\Theta_j}$ is a regular semigroup for each j. Let $\nu_{j,t}$ denote the standard measures of the semigroup on S_{Θ_j} and suppose that $B^{(j)}$ is the generating functional on $C^2_{lok}(S_{\Theta_j})$. With the notation of (7.4), $\mu_t = \sum_{j=1}^{r} \nu_{j,t}^{m_j}$. We remark that $B^{(j)}$ vanishes if Θ_j consists of a single point. Then we put $C^2_{lok}(S_{\Theta_j}) := C(\{1\})$.

(9.20) Theorem.

(a) Let $f \in C_b(S_k)$ denote a function such that $f \circ m_j$ belongs to $C^2_{lok}(S_{\Theta_j})$ for $j = 1,\dots,r$. Then denoting $\mu_o := \sum_{\theta \in \Theta} \varepsilon_{e_\theta}$ where $e_\theta := m_j(e_{\Theta_i})$, $\theta \in \Theta_j$,

$$\lim_{t\to 0} \frac{\int f\,d\mu_t - \int f\,d\mu_o}{t} = \sum_{j=1}^{r} B^{(j)}(f \circ m_j).$$

(b) The Hellinger transform has the form $H(E_t)(z) = H(E_{t\Theta_j})(m_j^{-1}(z))$ if $z \in m_j(S_{\Theta_j})$ and $H(E_t)(z) = 0$ otherwise.

Proof. The proof of (a) follows immediately from (7.4) and the corresponding theorems for regular experiments.

ad (b). Suppose that $z \notin \bigcup_{j=1}^{r} m_j(S_{\Theta_j})$. Then there are s,t in Θ such that $z_s > 0$, $z_t > 0$ but s,t do not lie in a common set Θ_j. If i is an arbitrary parameter (which belongs to one set Θ_j), then by (7.4) $p_i \cdot \mu_t$ is concentrated on $m_j(S_{\Theta_j})$. Hence, $f_z = 0$ $p_i \cdot \mu_t$ a.e., and the Hellinger transform vanishes. □

<u>(9.21) Example.</u> (Binary experiments) Let Θ consist of two points. Then $\varphi_2 \colon S_2 \longrightarrow [-\infty, \infty]$ can be extended if we put $\varphi_2((0,1)) := \infty$. If $E = (S_2, \mathcal{B}(S_2), (\mu_i)_{i=1,2})$ is a standard experiment and $\nu_i := \mu_i^{\varphi_2}$, then E and $F := ([-\infty,\infty], \mathcal{B}([-\infty,\infty]), (\nu_i)_{i=1,2})$ are equivalent. The measures ν_i have the following interpretation: ν_1 is the distribution of the log-likelihood ratio $\log \frac{d\mu_2}{d\mu_1}$ w.r.t. μ_1 and ν_2 is the distribution of $-\log \frac{d\mu_1}{d\mu_2}$ w.r.t. μ_2. Note that F is uniquely determined by $\nu_{1|\mathbb{R}^1}$. $(\nu_1(\{-\infty\}) = 1 - \nu_1(\mathbb{R}^1)$ and $\nu_2 = \exp(y) \cdot \nu_{1|\mathbb{R}^1} + (1 - \int \exp(y) d\nu_1(y))\,\varepsilon_\infty)$. F is infinitely divisible iff $\nu_{1|\mathbb{R}^1}$ is infinitely divisible on $\mathbb{R}^1$. In this case we can compute the decomposition $E \sim E_1 \otimes E_2$ of (9.11). If we put $a_1 := -\log(\nu_1(\mathbb{R}^1))$ and $a_2 := -\log(\int_{\mathbb{R}^1} \exp(y) d\nu_1(y))$ the Lévy measure has the form $\eta_2 := a_1\, \varepsilon_{-\infty} + a_2\, \varepsilon_\infty$.

On $[-\infty, +\infty]$ E_1 has the representation $E_1 \sim ([-\infty,\infty], \mathcal{B}([-\infty,\infty]), (\rho, \exp(y) \cdot \rho))$, where $\rho = \frac{1}{\nu_1(\mathbb{R}^1)} \nu_{1|\mathbb{R}^1} * \varepsilon_{a_2 - a_1}$. Moreover,

$$E_2 \sim ([-\infty,+\infty], \mathcal{B}([-\infty,+\infty]), (\exp(-a_1)\varepsilon_{a_1-a_2} + (1-\exp(-a_1))\varepsilon_{-\infty}, \\ \exp(-a_2)\varepsilon_{a_1-a_2} + (1-\exp(-a_2))\,\varepsilon_\infty)).$$

We remark that nonregular binary experiments consist of the measures $\varepsilon_{-\infty}$, ε_∞. Infinitely divisible binary experiments and their representation have been studied by LeCam, 1969, Ch. II. Binary experiments appear for example in the theory of testing. The concept of contiguity can be treated by the methods sketched in (9.21).

Finally, we show that generating functionals on S_k can be described in terms of almost positive functionals. We shall use the notation of (8.3) (viii) and put $X := S_k$, $x_o := e_k$, $V := C^2_{lok}(S_k)$.

<u>(9.22) Theorem.</u> Let $B: C^2_{lok}(S_k) \longrightarrow \mathbb{R}$ be a linear functional. Then the assertions (a) and (b) are equivalent:

(a) B is the generating functional of a continuous semigroup of regular experiments on S_k .

(b) (i) B is almost positive,

(ii) $B(p_i) = 0$ for each $i = 1,\dots,k$.

<u>Proof.</u> Generating functionals satisfy (b) (i), (ii) (see (9.15)).

b) $\Longrightarrow$ a). Suppose that V (and U) is a relative compact neighbourhood of e_k in S_k such that $U \subset \overline{U} \subset V \subset \overline{V} \subset G_k$. We consider a function $0 \le h \le 1$ in $C(S_k)$ vanishing on U and equal to 1 on CV . Then $f \longmapsto B(hf)$ is positive and bounded on $C(S_k)$ and there exists a bounded measure $\nu \ge 0$ on S_k such that $B(hf) = \int f\, d\nu$ and $\nu(U) = 0$. Let $(E_t^{(2)})_{t>0}$ denote the Poisson semigroup of experiments with the Lévy measure ν and B_2 the corresponding generating functional. We remark that $B_1(f) = B(f) - B_2(f)$ satisfies (b) (i), (ii). Note that $B_1(f) = B(f(1-h))$ if $f \ge f(e_k) = 0$. Moreover, $B_1(f) = 0$ if $f_{|V} = 0$.

Let $h' \in C_{oo}(G_k)$ satisfy $0 \le h' \le 1$, $h'_{|V} = 1$. For every $F \in C^2_{lok}(\mathbb{R}^{k-1})$ we may extend $h'(F \circ \varphi_k)$ continuously on S_k defining the extension to be 0 on $S_k \setminus G_k$. Then

$$F \longmapsto A(F) := B_1(p_1 h'(F \circ \varphi_k))$$

is an almost positive and tight functional on $C^2_{lok}(\mathbb{R}^{k-1})$ satisfying $A(1) = 0$. Hence, by (8.3) A defines a continuous semigroup $(\mu_t)_{t>0}$ in $M_1(\mathbb{R}^{k-1})$ such that the Lévy measure is concentrated on $\varphi_k(\overline{V})$. Then

$$A(g_i) = A\left(\frac{p_{i+1}}{p_1} \circ \varphi_k^{-1}{}_{|\mathbb{R}^{k-1}}\right) = B_1(p_{i+1} h') = 0 .$$

If we put $E_t^{(1)} := (\mathbb{R}^{k-1}, \mathcal{B}(\mathbb{R}^{k-1}), (\mu_t, g_1 \cdot \mu_t, \dots, g_{k-1} \cdot \mu_t))$, then $E_t^1 \otimes E_t^2$ has the generating functional B . □

Since S_k is compact, we do not need any tightness condition for B .

Let $EIR(\Theta)$ denote the set of classes of infinitely divisible regular experiments and D the corresponding convex cone of generating functionals. Then $\Phi : EIR(\Theta) \longrightarrow D$ is the bijection defined by $H(E)(z) =: \exp(\Phi(E)(f_z))$. We introduce the Hellinger transform $\hat{B}: S_k \longrightarrow \mathbb{R}^1$, $\hat{B}(z) := B(f_z)$, $B \in D$ and let T_D denote the topology of pointwise convergence of Hellinger transforms on D . Moreover, let S_D be the $\sigma(D, C^2_{lok}(S_k))$ - topology of pointwise convergence w.r.t. $C^2_{lok}(S_k)$ - functions on D . For every $f \in C^2_{lok}(S_k)$ we consider the extension $h = (f \circ \varphi_k^{-1}) \circ \varphi$ given before (9.15). If $x \in S_k$ put

$$\alpha(x) := \sum_{i=1}^{k} (x_i - \frac{1}{k})^2$$

and

$$\beta(x) := h(x) - h(e_k) - \sum_{i=1}^{k} (x_i - \frac{1}{k}) h_{x_i}(e_k) - \frac{1}{2} \sum_{i,j=1}^{k} (x_i - \frac{1}{k})(x_j - \frac{1}{k}) h_{x_i x_j}(e_k).$$

A Taylor expansion of h around e_k yields $\beta(x) = o(\alpha(x))$ as $x \longrightarrow e_k$. Hence,

$$\|f\|_0 = \sum_{i=1}^{k-1} |D_i f| + \frac{1}{2} \sum_{i,j=1}^{k-1} |D_i D_j f| + \|(\frac{\beta}{\alpha})_{|S_k}\|$$

is a norm on $C^2_{lok}(S_k)$ ($\|\cdot\|$ denotes the sup-norm).

<u>(9.23) Lemma.</u> Every $B \in D$ is $\|\cdot\|_0$- continuous.

<u>Proof.</u> The integral part of B has the form

$$\int_{S_k \setminus \{e_k\}} (h(x) - h(e_k) - \sum_{i=1}^{k} (x_i - \frac{1}{k}) h_{x_i}(e_k))\, d\eta(x)$$

$$= \frac{1}{2} \sum_{i,j=1}^{k} h_{x_i x_j}(e_k) \int (x_i - \frac{1}{k})(x_j - \frac{1}{k})\, d\eta(x) + \int (\frac{\beta}{\alpha}) \alpha\, d\eta .$$

Note that $|(x_i - \frac{1}{k})(x_j - \frac{1}{k})| \leq \alpha(x)$, $x \in S_k$; by (9.15) (iv) the measure $\alpha \cdot \eta$ is finite . □

Let $(X, \|\cdot\|_0)$ denote the completion of $(C^2_{lok}(S_k), \|\cdot\|_0)$. Then $D \subset X'$ carries the corresponding norm of operators. T_s denotes the topology of weak convergence of experiment-types.

<u>(9.24) Theorem.</u>

(a) $(EIR(\Theta), T_s)$ is locally compact.

(b) $\Phi : (EIR(\Theta), T_s) \longrightarrow (D, T_D)$ is a topological isomorphism.

(c) S_D is strictly finer than T_D .

(d) S_D and T_D coincide on all norm-bounded subsets of D .

Proof.

(a) $EIR(\Theta) = \{E \in EI(\Theta): H(E)(e_k) > 0\}$ is an open subset of the compact space $EI(\Theta)$.

(b) is well-known.

(c) Let us denote $V := C^2_{lok}(S_k)$. Suppose $S_D = T_D$. Then the $\sigma(X',V)$ - topology induced on D is locally compact. Since the difference of two compact sets remains compact, every point of the topological vector space $(D-D, \sigma(X',V))$ has a compact neighbourhood. The dimension of $D-D$ is finite which yields the desired contradition.

(d) W.l.g. let $A \subset D$ be a S_D closed and norm-bounded set. If we take into account (9.22) (b), then Alaoglu-Bourbaki's theorem implies that (A, S_D) is compact . □

Assertion (9.24) (d) implies the following convergence result.

(9.25) Corollary. Let $(E_n)_n$, E be regular infinitely divisible experiments for $\Theta = \{1,...,k\}$ having Lévy measures $(\eta_n)_n$, η . If $E_n \longrightarrow E$, weakly, then $(\eta_n)_n$ converges vaguely to η on $S_k \setminus \{e_k\}$.

Proof. 1. We remark that the set of Gaussian experiments (including the totally uninformative experiment) is closed in $EIR(\Theta)$.

2. Claim: $\overline{\lim}_{n\to\infty} \int \alpha \, d\eta_n < \infty$.

Let $\rho: x \longmapsto (x_1 - \frac{1}{k})^2$. We prove that $\overline{\lim}_{n\to\infty} \int \rho \, d\eta_n < \infty$. Let j be the index for which $|x_1 - x_i|$ is maximal. Then

$$\rho(x) = \frac{1}{k^2} \left(\sum_{i=2}^{k} (x_1 - x_i) \right)^2 \le (x_1 - x_j)^2$$

$$= \left(|\sqrt{x_1} - \sqrt{x_j}| \cdot |\sqrt{x_1} + \sqrt{x_j}| \right)^2 \le 4 \sum_{i=2}^{k} |\sqrt{x_1} - \sqrt{x_i}|^2$$

$$= 4 \sum_{i=2}^{k} (x_1 + x_i - 2\sqrt{x_1 x_i}) =: h(x), \quad x \in S_k .$$

Let B_n denote the generating functionals of E_n, $B_n^{(1)}$ the Gaussian part, and $B_n^{(2)}$ the integral term of B_n so that $B_n = B_n^{(1)} + B_n^{(2)}$, $n \in \mathbb{N}$. Weak convergence yields that $\lim\limits_{n\to\infty} B_n(h)$ exists. Applying (9.22) to $h-\rho \geq 0$ we obtain $B_n^{(1)}(\rho) \geq 0$ and $B_n(h) \geq B_n(\rho)$. Hence,

$$\overline{\lim_{n\to\infty}} \int \rho \, d\eta_n = \overline{\lim_{n\to\infty}} B_n^{(2)}(\rho) \leq \lim_{n\to\infty} B_n(h).$$

Let $E_n = E_n^{(1)} \otimes E_n^{(2)}$ denote the decomposition into a Gaussian experiment $E_n^{(1)}$ and a Poisson experiment (whose Hellinger transform is determined by the integral part of the Lévy-Khintchine formula). Suppose that B is the generating functional of E. Let $E_{k(n)}^{(2)} \longrightarrow F_2$ be a weak accumulation point. Then the Hellinger transforms show that $E_{k(n)}^{(1)}$ converges to a Gaussian experiment F_1 and $E \sim F_1 \otimes F_2$. Hence, F_2 is regular and admits a generating functional A_2. However, by the proof of (9.23) $(B_n^{(2)})_n$ is bounded. Applying (9.24) the sequence $B_{k(n)}^{(2)}$ converges to A_2 w.r.t. the $\mathcal{S}_D$-topology. Then for every $f \in C_{oo}(S_k \setminus \{e_k\})$

$$\int f \, d\eta = B(f) = A_2(f) = \lim_{n\to\infty} B_{k(n)}^{(2)}(f) = \lim_{n\to\infty} \int f \, d\eta_{k(n)}.$$

Finally, we note that for every subsequence of $\mathbb{N}$ there is another subsequence such that convergence is valid. □

(9.26) Remarks.

(a) If in addition the Gaussian part of the experiments $(E_n)_n$ is trivial, then (9.25) is contained in Lemma (4.6) and Corollary (4.8) of Part I.

(b) More details on almost positive operators are included in the paper of von Waldenfels, 1965. The author deals with compact spaces containing $\mathbb{R}^n$. This is just the situation satisfied for our simplex S_k.

We conclude this section with a discussion of the Lévy-Khintchine formula (9.15), (9.17). An experiment is Gaussian iff its Lévy measure vanishes or, equivalently, iff it is homogeneous and its corresponding measure in the sense of (9.12) is Gaussian. Then the matrix $(2c_{ij})$ is the covariance matrix and $(-c_{ii})$ is the mean of $(\log \frac{dP_i}{dP_1})_{i=1,\ldots,k}$ (w.r.t. P_1) which is a Gaussian process. If we change the basis of the log-likelihood process, this process remains Gaussian. If the matrix (c_{ij}) vanishes, then we speak of a Poisson experiment, see LeCam, 1974. We remark that Poisson experiments can be obtained as infinite products of compound Poisson experiments. Moreover, the Lévy-Khintchine representation shows that every regular infinitely divisible experiment can be decomposed as a product of a Gaussian experiment and a Poisson experiment (cf. also Part I, Sec. 5). We note that the analogous notion for infinitely divisible probability measures μ is a little bit different. If μ is an infinite convolution product of compound Poisson distributions then it is sometimes called a generalized Poisson distribution.

10. The Lévy-Khintchine Formula for Arbitrary Regular Infinitely Divisible Statistical Experiments

In this section we consider experiments $E = (X, \mathcal{A}, (P_\theta)_{\theta \in \Theta})$ with an arbitrary parameter set $\Theta \neq \emptyset$.

(10.1) Lemma.

(a) E is infinitely divisible iff the restriction E_I is infinitely divisible for every $I \in A(\Theta)$.

(b) $\mathcal{E}I(\Theta)$ is weakly compact.

(c) Every infinitely divisible experiment is embeddable into a continuous semigroup of experiments.

(d) Suppose that E is infinitely divisible. Then there is a unique partition $(\Theta_j)_{j \in Z}$ of Θ such that

(i) E_{Θ_j} is regular for every $j \in Z$.

(ii) P_r and P_s are mutually singular if r and s do not lie in a common set Θ_i.

Proof. (a). See Lemma (5.1) in Part I.

(b) - (d). The proofs of (9.8) (ii), (iv) and (9.19) carry over. □

To compare different experiments we introduce $S := \prod_{I \in A(\Theta)} S_I$, equipped with the product topology, and let $\xi_I : S \longrightarrow S_I$ denote the canonical projection.

(10.2) Lemma. For every experiment E there exists an equivalent experiment $F = (S, \mathcal{B}(S), (Q_\theta)_{\theta \in \Theta})$ such that each Q_θ is a Radon measure on S.

Proof. We consider the function $h: X \longrightarrow S$ given by

$$h(x) := \left(\left(\frac{dP_i}{d \sum_{j \in I} P_j} (x) \right)_{i \in I} \right)_{I \in A(\Theta)}$$

where $\left(\frac{dP_i}{d \sum_{j \in I} P_j}\right)$ denotes a fixed S_I - valued version of the Radon-Nikodym densities, $I \in A(\Theta)$. We define $Q_\theta := P_\theta^{\,h}$; this measure can be extended from $\bigotimes_{I \in A(\Theta)} \mathcal{B}(S_I)$ to $\mathcal{B}(S)$. We remark that

$$\left(\frac{dQ_i}{d \sum_{j \in I} Q_j}\right)_{i \in I} = \xi_I \quad \sum_{j \in I} Q_j \quad \text{a.e.},$$

and $E_I \sim F_I$ follows by considering Hellinger transforms. □

Note that Q_θ is not unique since h depends on the special version of the densities. For $I, J \in A(\Theta)$, $I \subset J$, put $N_{I,J} := \{x \in S_J : \sum_{i \in I} x_i = 0\}$; recall $A_i(\Theta) = \{I \in A(\Theta): i \in I\}$, $i \in \Theta$.

(10.3) Lemma. Let $i \in I$. Then

(i) $A_i = \{(x_J) \in S: \varphi_{I,J}(x_J) = x_I$ for every $x_J \notin N_{I,J}$; $I, J \in A_i(\Theta)$, $I \subset J\}$ is a closed subset of S.

(ii) Q_i is concentrated on A_i.

Proof. (i) Let $(x_\alpha)_\alpha \longrightarrow x \in S$ be a convergent net in A_i. Then $\xi_I(x_\alpha) \longrightarrow \xi_I(x)$. If $\xi_J(x) \in N_{I,J}$ there is nothing to prove. Otherwise, $S_J \setminus N_{I,J}$ is an open neighbourhood of $\xi_J(x)$ and $\xi_J(x_\alpha)$ lies in this set for $\alpha \geq \alpha_0$. Hence,

$$\xi_I(x) = \lim_\alpha \xi_I(x_\alpha) = \lim_\alpha \varphi_{I,J}(\xi_J(x_\alpha)) = \varphi_{I,J}(\xi_J(x)).$$

(ii) $A_i = \bigcap_{Z \in A_i(\Theta)} B_Z$ where

$$B_Z := \{(x_J)_J \in S: \varphi_{I,J}(x_J) = x_I \text{ for every } x_J \notin N_{I,J},\ I,J \in A_i(Z),\ I \subset J\}.$$

Note that B_Z is closed and the system $(B_Z)_Z$ is filtering to the left. Since Q_i is a Radon measure it is sufficient to prove $Q_i(B_Z) = 1$ for every B_Z.

For fixed $Z \in A_i(\Theta)$ we choose $I, J \in A_i(Z)$, $I \subset J$. Using the notation of the proof of (10.2), $\varphi_{I,J} \circ \xi_J \circ h = \xi_I \circ h$ P_i a.e. If we now take all possible pairs I,J in $A_i(Z)$, then $Q_i(B_Z) = 1$ follows. □

(10.4) Definition.

(a) $Y_i := \{(x_J)_J \in \prod_{J \in A_i(\Theta)} S_J : \varphi_{I,J}(x_J) = x_I$ for all $x_J \notin N_{I,J}$, $I,J \in A_i(\Theta)$, $I \subset J\}$.

(b) Let $\tau_i : A_i \longrightarrow Y_i$, $\pi_i^I : Y_i \longrightarrow S_I$ denote the canonical projections for $i \in \Theta$, $I \in A_i(\Theta)$.

We note that Y_i is a compact subspace of $\prod_{J \in A_i(\Theta)} S_J$.

(10.5) Lemma. Let $E_r = (X_r, A_r, (P_\theta^r)_{\theta \in \Theta})$, $r = 1,2$, be experiments having the representation $F_r = (S, \mathcal{B}(S), (Q_\theta^r)_{\theta \in \Theta})$ on S. The following assertions are equivalent:

(i) $E_1 \sim E_2$.

(ii) $(Q_i^1)^{\tau_i} = (Q_i^2)^{\tau_i}$ for every $i \in \Theta$.

Proof. (i) $\Longrightarrow$ (ii). Let $\mathcal{A}$ denote the set of functions $f \in C(Y_i)$ having the following properties: There exists a set $I \in A_i(\Theta)$ and a function $g \in C_b(T_i^I)$ such that $f(x) = g(\pi_i^I(x))$ for every $x \in (\pi_i^I)^{-1}(T_i^I)$. We show that $\mathcal{A}$ is an algebra in $C(Y_i)$ which contains the constant functions and separates the points of Y_i. Consider $f_r \in \mathcal{A}$, $r = 1,2$, such that

$$f_r(x) = g_r \circ \pi_i^{I_r}(x), \quad x \in (\pi_i^{I_r})^{-1}(T_i^{I_r}).$$

Put $I := I_1 \cup I_2$ and $\tilde{g}_r := g_r \circ \varphi_{I_r,I|T_i^I}$. Note that $\pi_i^{I_r}(x) \in T_i^{I_r}$ and $\pi_i^{I_r}(x) = \varphi_{I_r,I} \circ \pi_i^I(x)$, $x \in (\pi_i^I)^{-1}(T_i^I)$. Hence, $f_r(x) = \tilde{g}_r \circ \pi_i^I(x)$ for these points x. Therefore $\mathcal{A}$ is an algebra. Two different points $x,y \in Y_i$ have different coordinates $\pi_i^I(x)$, $\pi_i^I(y)$ for some $I \in A_i(\Theta)$. These coordinates can be separated by a suitable function $g \in C_b(S_I)$. Then $g \circ \pi_i^I \in \mathcal{A}$ separates x and y. Now, the Stone-Weierstrass theorem implies that the Radon measures defined in (ii) coincide if

$$\int f \circ \tau_i \, dQ_i^1 = \int f \circ \tau_i \, dQ_i^2 \quad \text{for every } f \in \mathcal{A}.$$

If $f \in \mathcal{A}$ has the representation $f = g \circ \pi_i^I$ on $(\pi_i^I)^{-1}(T_i^I)$, then

$\int f \circ \tau_i \, dQ_i^r = \int g\left(\left(\frac{dP_j^r}{d \sum_{n \in I} P_n^r}\right)_{j \in I}\right) dP_i^r$, $r = 1,2$. Since $(E_1)_I$ and $(E_2)_I$ are equivalent, the latter integrals coincide for $r = 1,2$.

(ii) $\Longrightarrow$ (i). Consider $I \in A_i(\Theta)$ and $f = g \circ \pi_i^I$ $(g \in C(S_I))$. Then Assumption (ii) and the above calculation imply

$$P_i^1 {}^{\left(\frac{dP_j^1}{d \sum_{n \in I} P_n^1}\right)_{j \in I}} = P_i^2 {}^{\left(\frac{dP_j^2}{d \sum_{n \in I} P_n^2}\right)_{j \in I}}$$

Therefore, the standard experiments of $(E_1)_I$ and $(E_2)_I$ coincide. □

If we compare these results with the case of a finite parameter set considered in § 9, then the simplex S_Θ is substituted by S and T_i^Θ by Y_i.

<u>(10.6) Lemma.</u> Let $(E_t = (X_t, A_t, (P_{i,t})_{i \in \Theta}))_{t > 0}$ be a family of experiments and $(F_t = (S, \mathcal{B}(S), (\mu_{i,t})_{i \in \Theta}))_{t > 0}$ be equivalent representations on S. The subsequent assertions are equivalent.

(i) $(E_t)_{t > 0}$ is a continuous semigroup of experiments.

(ii) For every $i \in \Theta$ and every $I \in A_i(\Theta)$ the family $(\mu_{i,t}^{\pi_i^I \circ \tau_i})$ is a continuous convolution semigroup on T_i^I.

This is a consequence of (9.8). In the following we need a notation which is independent of the numerical ordering of the finite parameter set $\{1, \ldots, k\}$.

<u>(10.7) Definition.</u>

(a) $\varphi_I^i : T_i^I :\longrightarrow [-\infty, \infty)^{I \setminus \{i\}}$, $\varphi_I^i(x) := \left(\log \frac{x_j}{x_i}\right)_{j \in I, j \neq i}$.

(b) $\mathcal{D}(Y_i)$ denotes the following set of functions $f \in C(Y_i)$: There exists a set $I \in A_i(\Theta)$, $|I| \geq 2$, and a function $g \in C_b(T_i^I)$ such that

(i) $f(x) = g(\pi_i^I(x))$ for every $x \in (\pi_i^I)^{-1}(T_i^I)$ and

(ii) $g \circ (\varphi_I^i)^{-1}|_{\mathbb{R}^{I \setminus \{i\}}}$ is a $C^2_{lok}(\mathbb{R}^{I \setminus \{i\}})$ function.

(c) If $f \in \mathcal{D}(Y_i)$ and $j \in I$ then $D_j^i f$ is the jth partial derivative of $g \circ (\varphi_I^i)^{-1}$ at $0 \in \mathbb{R}^{I \setminus \{i\}}$. If $j \notin I$ put $D_j^i f = 0$. $D_j^i D_k^i f$ denotes the second partial derivative of $g \circ (\varphi_I^i)^{-1}$ at the origin w.r.t. j and k if $\{j,k\} \subset I$ and $D_j^i D_j^i f = 0$ otherwise.

(10.8) Lemma. For every $z \in T_j^J$ there exists a $x \in Y_i$ such that $\pi_i^J(x) = z$ if $j \in J \in A_i(\Theta)$.

Proof. If $Z \in A_i(J)$ we define the injection $m_Z: S_Z \longrightarrow S_J$ by $p_\theta^J(m_Z(y)) := p_\theta^Z(y)$, $\theta \in Z$, and $p_\theta^J(m_Z(y)) := 0$, otherwise. Put $J_o := \{\theta \in J: p_\theta^J(z) = 0\}$.

1. Suppose $i \notin J_o$. Then we define $x = (x_I)_{I \in A_i(\Theta)}$ by $x_I := m_{I \cap J}(\varphi_{I \cap J, J}(z))$.

2. If $i \in J_o$ we first define x_I for $I \subset J$. Put $x_I := e_I$ if $I \subset J_o$ and $x_I := \varphi_{I,J}(z)$ if $I \not\subset J_o$. In the general case put $x_I := m_{I \cap J}(x_{I \cap J})$. We show that $x \in Y_i$. For this we consider $I_1, I_2, I_1 \subset I_2$ in $A_i(\Theta)$. We prove $\varphi_{I_1,I_2}(x_{I_2}) = x_{I_1}$ whenever $x_{I_2} \notin N_{I_1,I_2}$. First, suppose $I_2 \cap J \subset J_o$; then the conclusion is trivial. If $I_1 \cap J \not\subset J_o$ then $I_2 \cap J \not\subset J_o$; hence, $\varphi_{I_1,I_2}(x_{I_2}) = x_{I_1}$ by construction of x. Now, suppose $I_2 \cap J \not\subset J_o$ but $I_1 \cap J \subset J_o$. Note that $x_{I_2} = m_{I_2 \cap J}(x_{I_2 \cap J})$. Since $I_1 \cap J \subset J_o$, the coordinate x_{I_2} lies in N_{I_1,I_2} and there is nothing to prove. □

(10.9) Lemma.

(a) $\mathcal{D}(Y_i)$ is dense in $C(Y_i)$ w.r.t. the sup-norm.

(b) The derivatives $D_j^i f$ and $D_j^i D_k^i f$ are independent of the particular representation $f = g \circ \pi_i^I$ of $f \in \mathcal{D}(Y_i)$.

Proof. (a). This is shown in the first part of the proof of (10.5).

(b). Suppose that f has the form $f = g \circ \pi_i^I$ on $(\pi_i^I)^{-1}(T_i^I)$ and

$f = \tilde{g} \circ \pi_i^J$ on $(\pi_i^J)^{-1}(T_i^J)$ and $I, J \in A_i(\Theta)$, $I \subset J$. Then by (10.8): $\tilde{g}(z) = g \circ \varphi_{I,J}(z)$ for every $z \in T_i^J$ and $\tilde{g} \circ (\varphi_J^i)^{-1} = (g \circ (\varphi_I^i)^{-1}) \circ \varphi_I^i \circ \varphi_{I,J} \circ (\varphi_J^i)^{-1}$. Observe that $\varphi_I^i \circ \varphi_{I,J} \circ (\varphi_J^i)^{-1}$ is the canonical projection of $\mathbb{R}^{J \setminus \{i\}}$ onto $\mathbb{R}^{I \setminus \{i\}}$. □

(10.10) Definition. If $f \in \mathcal{D}(Y_i)$ then $I(f) := \{j \in \Theta : D_j^i f \neq 0\} \cup \{i\}$. We define a measurable function Γf on Y_i: If $I(f) = \{i\}$ we put $\Gamma f := 0$; otherwise,

$$\Gamma f(x) := \begin{cases} \sum_{j \in I(f) \setminus \{i\}} D_j^i f \left(\dfrac{p_j^{I(f)}(\pi_i^{I(f)}(x))}{p_i^{I(f)}(\pi_i^{I(f)}(x))} - 1 \right), & x \in (\pi_i^{I(f)})^{-1}(T_i^{I(f)}) \\ 0, & \text{otherwise}. \end{cases}$$

(10.11) Lemma. Suppose that $f \in \mathcal{D}(Y_i)$ has the form $f(x) = g \circ \pi_i^J(x)$ for every $x \in (\pi_i^J)^{-1}(T_i^J)$. Then

$$\Gamma f(x) = \sum_{j \in J \setminus \{i\}} D_j^i f \left(\frac{p_j^J(\pi_i^J(x))}{p_i^J(\pi_i^J(x))} - 1 \right), \quad x \in (\pi_i^J)^{-1}(T_i^J).$$

Proof. We remark $I(f) \subset J$. Since $\varphi_{I(f),J}(\pi_i^J(x)) = \pi_i^{I(f)}(x)$ the proof follows from (10.9) (b). □

By (10.6) $(\mu_{i,t}^{\tau_i})_{t>0}$ can be regarded as a "projective limit" of convolution semigroups on Y_i. As in the finite case the proof of the general Lévy-Khintchine formula for experiments uses the representation of semigroups defined by one parameter i. Let us denote by $r(i)$ the element $(e_I)_{I \in A_i(\Theta)}$ in Y_i.

(10.12) Theorem. Let $(E_t)_{t>0}$ be a regular continuous semigroup of experiments, $i \in \Theta$, and $(\mu_{i,t}^{\tau_i})_{t>0}$ denote the corresponding distributions on Y_i (cf. (10.6)). Then:

(a) For every $f \in C_{oo}(Y_i \setminus \{r(i)\})$ the limit

$$\lim_{t \to 0} \frac{\int f \circ \tau_i \, d\mu_{i,t}}{t} =: \int f \, d\eta_i$$

exists ; η_i is a positive Radon measure on $Y_i \setminus \{r(i)\}$ satisfying

(i) $\displaystyle\int (p_i^I(\pi_i^I(x)))^{-1} \sum_{j \in I} (p_j^I(\pi_i^I(x)) - \frac{1}{|I|})^2 \, d\eta_i(x) < \infty , \quad I \in A_i(\Theta).$

(b) For every $f \in \mathcal{D}(Y_i)$ the limit

$$B_i(f) := \lim_{t \to 0} \frac{\int f \circ \tau_i \, d\mu_{i,t} - f(r(i))}{t}$$

exists ; B_i has the representation

$$B_i(f) = \sum_{j \in \Theta \setminus \{i\}} b_j D_j^i f + \sum_{j,k \in \Theta \setminus \{i\}} a_{jk} D_j^i D_k^i f$$

$$+ \int_{Y_i \setminus \{r(i)\}} (f - f(e(i)) - \Gamma f) \, d\eta_i .$$

The measure η_i is called the Lévy measure and B_i the generating functional of $(\mu_{i,t}^{\tau_i})_{t > 0}$. Moreover,

(ii) $(a_{jk})_{j,k \in \Theta \setminus \{i\}}$ is real and positive semi-definite,

(iii) $b_j = -a_{jj} - \eta_j(\{p_i^{\{i,j\}} \circ \pi_j^{\{i,j\}} = 0\})$ for $j \neq i$.

(Note that η_j lies on $Y_j \setminus \{r(j)\}$.) The representation of B_i is unique.

(c) Let η_I denote the Lévy measure of the semigroup $(E_t)_I$ for $I \in A_i(\Theta)$, $|I| \geq 2$ (cf. (9.15)). Then for every $h \in C_{oo}(S_I \setminus \{e_I\})$

$$\int h \circ \pi_i^I \, d\eta_i = \int h \, p_i^I \, d\eta_I \quad \text{and} \quad \sum_{j \in I} \int h \circ \pi_j^I \, d\eta_j = \int h \, d\eta_I .$$

Sometimes it is useful to know that all Lévy measures can be defined on a common space. Therefore we first prove the next lemma.

(10.13) Lemma. Let $i \in \Theta$. Then there is a Radon measure η_i on $Y_i^* := Y_i \setminus \{r(i)\}$ such that for every $I \in A_i(\Theta)$ and every $h \in C_{oo}(S_I \setminus \{e_I\})$

(i) $\displaystyle\int h \circ \pi_i^I \, d\eta_i = \int h \, p_i^I \, d\eta_I , \quad I \in A_i(\Theta) .$

<u>Proof.</u> Let V_i denote the following subspace of functions of $C_{oo}(Y_i^*)$: There exist a set $I \in A_i(\Theta)$ and a function $g \in C_b(T_i^I)$ such that $f(x) = g \circ \pi_i^I(x)$ for every $x \in (\pi_i^I)^{-1}(T_i^I) \setminus \{r(i)\}$. We show:

1. V_i is positively rich in the sense of Bourbaki, 1965, p. 55, i.e.: For every compact $K \subset Y_i^*$ there is a relatively compact neighbourhood U of K such that for all $\varepsilon > 0$ and all positive $f \in C_{oo}(Y_i^*)$ supported by K there exists a positive $h \in V_i$ such that the sup-norm satisfies $\|h - f\| \leq \varepsilon$ and the support of h lies in U. Moreover, V_i is a linear subspace.

2. If $f \in V_i$ we can choose I large enough such that $f(x) = g \circ \pi_i^I(x)$ for all $x \in (\pi_i^I)^{-1}(T_i^I) \setminus \{r(i)\}$ and g vanishes in some neighbourhood of $e_I \in S_I$.

1) Let $\mathcal{A}$ be the dense subalgebra of $C(Y_i)$ introduced in the proof of (10.5). Then there is a function $h' \in \mathcal{A}$ of the form $h' = g' \circ \pi_i^I$ on $(\pi_i^I)^{-1}(T_i^I)$ such that $\|f - h'\| \leq \varepsilon/2$ if f is a positive function in $C_{oo}(Y_i^*)$ supported by K. Choose $\delta\colon \mathbb{R} \longrightarrow \mathbb{R}$, $\delta(t) := \max(0, t - \varepsilon/2)$ and introduce $h := \delta \circ h'$, $g := \delta \circ g'$. The support of h is contained in K and $\|f - h\| \leq \varepsilon$ since $\|h - h'\| \leq \varepsilon/2$. Moreover, $0 \leq h \in V_i$. The arguments used in the proof of (10.5) imply that V_i is a linear subspace of $C_{oo}(Y_i^*)$.

2) Suppose $f \in V_i$, $f = g \circ \pi_i^I$ on $(\pi_i^I)^{-1}(T_i^I) \setminus \{r(i)\}$, $I \in A_i(\Theta)$, and let $K \subset Y_i^*$ denote the compact support of f. Then there is a neighbourhood $V \subset Y_i$ of $r(i)$ such that $K \cap V = \emptyset$. We can choose a finite subset $\{J_1,\ldots,J_n\}$ of $A_i(\Theta)$, a family of open neighbourhoods $U_{J_k} \subset G_{J_k}$ of e_{J_k}, $k = 1,\ldots,n$, such that

$$(U_{J_1} \times \ldots \times U_{J_n} \times \prod_{\substack{J \in A_i(\Theta) \\ J \neq J_k}} S_J) \cap Y_i \subset V .$$

Choose $J = I \cup \bigcup_{k=1}^{n} J_k$ and put $U_J := \bigcap_{k=1}^{n} \varphi_{J_k,J}^{-1}(U_{J_k}) \cap G_J$. Then $U_J \subset S_J$ is an open neighbourhood of e_J and $\pi_i^J(K) \cap U_J = \emptyset$, since $(\pi_i^J)^{-1}(U_J) \subset V$. Consequently, we may put $g' := g \circ \varphi_{I,J}$, and $f(x) = g' \circ \pi_i^J(x)$ follows for every $x \in (\pi_i^J)^{-1}(T_i^J) \setminus \{r(i)\}$. By (10.8) for every $z \in T_i^J$ there is a $x \in (\pi_i^J)^{-1}(T_i^J)$ such that $\pi_i^J(x) = z$. Hence, g' vanishes on U_J and statement 2. is proved.

3) Let $\mu_{i,t}^{I}$ ($:= \mu_{i,t}^{\pi_i^I \circ \tau_i}$) denote the ith convolution semigroup of the semigroup $(E_t)_I$, $I \in A_i(\Theta)$. Then for all $I,J \in A_i(\Theta)$, $I \subset J$,

$\mu_{i,t}^{I} = (\mu_{i,t}^{J})^{\varphi_{I,J}}$:

For all $f \in V_i$ having the form 2. we compute

$$N(f) = \lim_{t \to 0} \frac{\int f \circ \tau_i \, d\mu_{i,t}}{t} = \lim_{t \to 0} \frac{\int g \, d\mu_{i,t}^{I}}{t} = \int g \, p_i^I \, d\eta_I$$

if we take into account (9.14). We note that $N(f)$ does not depend on the choice of the basis I since $(\mu_{i,t}^{I})_I$ is a "projective system" . Moreover, N is linear and positive on V_i . Since V_i is positively rich, N has a unique extension to a Radon measure η_i on Y_i^* , Bourbaki, 1965, Prop. 2, p. 56 . □

<u>Proof of Theorem (10.12).</u> a) We prove convergence for all $f \in C_{oo}(Y_i^*)$. For a relatively compact neighbourhood U of the support of f we choose another relatively compact neighbourhood $W \subset Y_i^*$ containing the closure of U . Choose a $C_{oo}(Y_i^*)$-map f' satisfying $0 \leq f' \leq 1$, $f'|_U = 1$, $f'|_{CW} = 0$. Then there is a non-negative $h \in V_i$ such that $\|f' - h\| \leq 1/2$. Hence, $2h \geq 1_U$. We note that $\eta_i|_U$ is finite and $\overline{\lim_{t \to 0}} \frac{1}{t} (\mu_{i,t}(U)) < \infty$ since (a) has been proved for h in (10.13). Finally, we take a sequence of functions h_n in V_i such that $\|h_n - f\| \longrightarrow 0$ and all supports of the h_n are contained in U . Then the triangular inequality proves the statement. (i) is a consequence of (9.15) (iv).

b) Let $f \in \mathcal{D}(Y_i)$ have the form (10.7) (i), (ii). Then $\int f \circ \tau_i \, d\mu_{i,t} = \int g \, d\mu_{i,t}^{I}$ and $B_i(f)$ exists in view of (9.14). For fixed $I \in A_i(\Theta)$ $B_i(f)$ admits the representation given in (9.14):

$$B_i(f) = \sum_{j \in I \setminus \{i\}} b_j D_j^i f + \sum_{j,k \in I \setminus \{i\}} a_{jk} D_j^i D_k^i f + \int_{Y_i^*} (f - f(r(i)) - \Gamma f) \, d\eta_i$$

where the matrix $(a_{jk})_{j,k \in I \setminus \{i\}}$ and the vector $(b_j)_{j \in I \setminus \{i\}}$ are defined in (9.14) (cf. (10.12)). Applying Lemma (7.2) and Part 3) of the previous proof we remark that $B_i(f)$ does not depend on the representation of f (cf. (10.11)).

Note that the Lévy measure η_j on Y_j^* fulfils the relation

$$\eta_j|_{(\pi_j^I)^{-1}(C\{e_I\})}^{\pi_j^I} = p_j^I \cdot \eta_I .$$

On the other hand the Lévy measures η_I are connected by

$$p_j^I \cdot \eta_I = (p_j^J \cdot \eta_J)_{|\varphi_{I,J}^{-1}(C\{e_I\})}^{\varphi_{I,J}} \quad \text{if} \quad I \subset J, \ I,J \in A_j(\Theta) .$$

Hence, $(p_j^I \cdot \eta_I)(\{p_i^I = 0\}) = \eta_j(\{p_i^{\{i,j\}} \circ \pi_j^{\{i,j\}} = 0\})$ if $\{i,j\} \subset I$. Now, we show that the representation is independent of the choice of I. For this we denote by B_i^I the generating functionals of the semigroups $(\mu_{i,t}^I)$ introduced in the proof of (10.13). We choose $I,J \in A_i(\Theta)$, $|I| \geq 2$, $I \subset J$ and consider $f \in C^2_{lok}(T_i^I)$. Then $f \circ \varphi_{I,J}|_{T_i^J} \in C^2_{lok}(T_i^J)$ and $B_i^I(f) = B_i^J(f \circ \varphi_{I,J})$. Note that the derivatives of $f \circ \varphi_{I,J}$ vanish for parameters in $J \setminus I$. Therefore the covariance matrix of the representation of B_i^I coincides with the restriction of the corresponding matrix of B_i^J to $I \setminus \{i\}$ since the representation of B_i^I is unique. □

We give a condensed formula for Hellinger transforms. Let f_z^I denote the function f_z on S_I introduced in (9.5).

<u>(10.14) Corollary.</u> (Lévy-Khintchine formula) Suppose that the assumptions of (10.12) are satisfied. Let $i_o \in \Theta$.

(a) For every $t > 0$ $H(E_t)$ has the form $H(E_t)(z) = \exp(t\Psi(z))$. The function

$$\Psi : \bigcup_{I \in A(\Theta)} S_I \longrightarrow (-\infty, 0]$$

admits the representation

$$\Psi(z) := \sum_{i,j \in I} c_{ij} z_i z_j - \sum_{i \in I} c_{ii} z_i + \sum_{i \in I} \int_{Y_i \setminus \{r(i)\}} (f_z^I \circ \pi_i^I - \sum_{j \in I} z_j p_j^I \circ \pi_i^I) \, d\eta_i ,$$

$$z = (z_i) \in S_I .$$

(i) $(\eta_i)_{i \in \Theta}$ is a family of Lévy measures on $(Y_i \setminus \{r(i)\})_{i \in \Theta}$ such that (10.12) (i) and the following statement (ii) are satisfied:

(ii) For every $I \in A(\Theta)$, $i \in I$, $f \in C_{oo}(S_I \setminus \{e_I\})$

$$\sum_{j \in I} \int (f \circ \pi_j^I)(p_i^I \circ \pi_j^I)\, d\eta_j = \int f \circ \pi_i^I \, d\eta_i .$$

(iii) $(c_{ij})_{i,j \in \Theta \setminus \{i_o\}}$ is real, positive semi-definite and

$$c_{i_o j} = c_{i i_o} = 0 \quad \text{for every } \ i,j \in \Theta .$$

(iv) By (i) - (iii) $(c_{i,j})$, η_i are uniquely determined.

(b) Assume that a pair $((c_{ij})_{i,j \in \Theta}, (\eta_i)_{i \in \Theta})$ satisfies (i) - (iii). Then $\exp(t\Psi)$ defines the Hellinger transforms of a continuous semigroup of regular experiments.

Proof. ad (a). Since E_t is regular $\Psi > -\infty$ follows. W.l.g. we may assume that $i_o \in I$ if $z \in S_I$. (Otherwise consider $J = I \cup \{i_o\}$ and $w \in S_J$ such that $\varphi_{I,J}(w) = z$ and $w_{i_o} = 0$.) If (a_{ij}) is the matrix of B_{i_o} ((10.12)) we put $c_{ij} := 0$ if i or $j = i_o$ and $c_{ij} := a_{ij}$, otherwise. Then (9.17) and (10.12) c) prove that Ψ has the desired form and (iii) is valid. If we put $h := f \cdot p_i^I$ then (ii) can be derived. The uniqueness is obvious (cf. (10.12)).

ad (b). Suppose that the assumptions of (b) are satisfied. Then for $I \in A_{i_o}(\Theta)$, $|I| \geq 2$ we define the Radon measure η_I by

$$\int h \, d\eta_I := \sum_{j \in I} \int h \circ \pi_j^I \, d\eta_j , \quad h \in C_{oo}(S_I \setminus \{e_I\}).$$

Applying (ii) to $h := f \cdot p_i^I$ we conclude

$$\int f \, p_i^I \, d\eta_I = \int f \circ \pi_i^I \, d\eta_i .$$

We consider the function f_i which vanishes on CT_i^I and is on T_i^I defined by $f_i(x) := (p_i^I(x))^{-1} \sum_{j \in I} (x_j - \frac{1}{|I|})^2$. Theorem (10.12) yields $\int f_i \, p_i^I \, d\eta_I < \infty$. Hence, observing

$$\sum_{j \in I} (x_j - \frac{1}{|I|})^2 \leq \sum_{j \in I} f_i(x) p_i^I(x) ,$$

η_I is a Lévy measure on S_I. Our assumptions imply that for $z \in S_I$ Ψ has the form

$$\Psi(z) = \sum_{i,j \in I} c_{ij}\, z_i\, z_j - \sum_{i \in I} c_{ii}\, z_i$$
$$+ \int (f_z^I - \sum_{j \in I} z_j\, p_J^I)\, d\eta_I$$

and $\exp(t\Psi)$ gives the Hellinger transforms of a continuous regular semigroup of experiments for I. Moreover, for fixed t a compatible family of experiments is defined. Now, (10.1) proves the result. □

In the following it will be shown that all Lévy measures can be defined on a common space.

(10.15) Notation. Let $\mathcal{R}$ be the ring

$$\mathcal{R} := \{A \in \bigotimes_{I \in A(\Theta)} \mathcal{B}(S_I) : \complement A \text{ is a neighbourhood of } (e_I)_I \in S\}$$

on S. For $I, J \in A(\Theta)$, $I \subset J$, define

$$\varphi_{I,J}^{-1}(A) := \{x \in S_J \setminus N_{I,J} : \varphi_{I,J}(x) \in A\}, \quad A \subset S_I,$$
$$L_I := S \setminus \{\xi_T = e_T : \text{ for all } T \subset I\}.$$

Let $\gamma_j : Y_j \longrightarrow S$ be the map defined by $\gamma_j((x_J)_{J \in A_j(\Theta)}) = (y_I)_{I \in A(\Theta)}$,

$$y_I := \begin{cases} x_I & j \in I \\ \varphi_{I, I \cup \{j\}}(x_{I \cup \{j\}}) & x_{I \cup \{j\}} \notin N_{I, I \cup \{j\}} \\ e_I & \text{otherwise} \end{cases}$$

then γ_j is $(\mathcal{B}(Y_j), \bigotimes_{I \in A(\Theta)} \mathcal{B}(S_I))$ - measurable $(j \in \Theta)$.

In the sequel let E be a regular infinitely divisible experiment for Θ and $(\eta_I)_{I \in A(\Theta)}$ its projective system of Lévy-measures. Recall that for every projective system of Lévy-measures in the sense of Part I there exists a regular E such that η_I is the Lévy measure of E_I, $I \in A(\Theta)$.

(10.16) Lemma. There exists a family of measures $(m_t)_{t \in \Theta}$ on $(S, \sigma(\mathcal{R}))$ such that

(i) $m_t(R) < \infty$, $R \in \mathcal{R}$, $t \in \Theta$;

(ii) if $h \in C_{oo}(S_I \setminus \{e_I\})$, $I \in A(\Theta)$, $t \in I$, then

$\int h \circ \xi_I \, dm_t = \int h \, p_t^I \, d\eta_I$ and

(iii) $\int h \circ \xi_I \, d \sum_{t \in I} m_t = \int h \, d\eta_I$

(iv) Let $T_1, T_2 \in A(\Theta)$, $T_1 \subset T_2$. Then

$$\left(\frac{dm_t|_{L_{T_2}} \cap \sigma(\xi_I: I \subset T_2)}{d \sum_{s \in T_1} m_s|_{L_{T_2}} \cap \sigma(\xi_I: I \subset T_2)} \right)_{t \in T_1} = \varphi_{T_1,T_2} \circ \xi_{T_2}|L_{T_2}$$

$\sum_{s \in T_1} m_s$ a.e.

The proof is divided into several steps.

1. Definition of the m_t. By (10.13) there is an extension of η_t to an abstract measure on $(Y_t, \mathcal{B}(Y_t))$ satisfying $\eta_t(\{r(e(t))\}) = 0$. Let m_t be the image measure $m_t := \eta_t^{\gamma_t}|_{\sigma(\mathcal{R})}$.

2. Proof of (i). Let $A \in \mathcal{R}$. First we assume that $A = CU_I \times \prod_{J \neq I} S_J$ where U_I is an open neighbourhood of $e_I \in S_I$. Since $e_I \notin CU_I$, the definition of γ_j yields

$$(1) \quad \gamma_j^{-1}(A) = Y_j \cap (\varphi_{I,I \cup \{j\}}^{-1}(CU_I) \times \prod_{J \in A_j(\Theta) \setminus \{I \cup \{j\}\}} S_J).$$

Thus, $\eta_j(\gamma_j^{-1}(A)) < \infty$ ($\varphi_{I,I \cup \{j\}}^{-1}(CU_I)$ is the complement of a neighbourhood of $e_{I \cup \{j\}} \in S_{I \cup \{j\}}$).

In general, if U_{I_k} are open neighbourhoods of e_{I_k} and $CA \supset U_{I_1} \times \ldots \times U_{I_n} \times \prod_{J \neq J_k} S_J$, then

$$\eta_j(\gamma_j^{-1}(A)) \leq \eta_j(\bigcup_{k=1}^{n} CU_{F_k} \times \prod_{J \neq J_k} S_J) < \infty.$$

3. (ii), (iii) hold. Note that

$$(2) \quad \xi_I \circ \gamma_j = \pi_j^I \quad \text{if} \quad j \in I.$$

Hence,

$$\int h \circ \xi_I \, dm_t = \int_{Y_t} h \circ \xi_I \circ \gamma_t \, d\eta_t$$
$$= \int h \circ \pi_t^I \, d\eta_t = \int h \circ p_t^I \, d\eta_I, \; t \in I$$

(cf. (10.13)). In order to prove (iv) we need the following lemma.

<u>(10.17) Lemma.</u> Let $A_k = (A_{I_k} \times \prod_{I \neq I_k, I \in A_t(\theta)} S_I) \cap Y_t$, $k = 1,\dots,n$, $I_k \in A_t(\theta)$, be such that CA_{I_k} is a measurable neighbourhood of e_{I_k} for each k. If $T \in A_t(\theta)$ and $\bigcup_{k=1}^{n} I_k \subset T$, then

$$\eta_t(\bigcap_{k=1}^{n} A_k \,\Delta(\bigcap_{k=1}^{n} \varphi_{I_k,T}^{-1}(A_{I_k}) \times \prod_{J \in A_t(\theta) \setminus \{T\}} S_J) \cap Y_t) = 0 .$$

<u>Proof.</u> By (10.13) $\eta_t((\pi_t^T)^{-1}(N_{I_k,T})) = 0$. Put $g_k := 1_{A_{I_k}}$; then

$$g_k \circ \varphi_{I_k,T} \circ \pi_t^T = g_k \circ \pi_t^{I_k} \quad \eta_t \text{ a.e.}, \quad 1 \le k \le n .$$

Considering the product $g_1 \cdot \ldots \cdot g_n$ the result follows. □

<u>Proof of (10.16) (iv).</u>

4. Define $R_I := R \cap \sigma(\xi_J : J \subset I)$, $I \in A(\theta)$. Then it is well-known that two measures μ_1, μ_2 on $(L_I, \sigma(R_I) \cap L_I)$ coincide if

(a) $\mu_i(R) < \infty$, $R \in R_I$, $i = 1,2$, and

(b) $\mu_1(R) = \mu_2(R)$ for all $R \in R_I$ such that

(3) $R = A_{J_1} \times \ldots \times A_{J_n} \times \prod_{J \neq J_k} S_J$, $J_k \subset I$,

and CA_{J_k} is a measurable neighbourhood of

$e_{J_k} \in S_{J_k}$, $k = 1, \dots, n$, $n \ge 1$

5. First, consider the case $T = T_1 = T_2$. Then it is sufficient to prove

(4) $$\int_R p_t^T \circ \xi_T \, d \sum_{s \in T} m_s = \int_R d\, m_t$$

for R being defined in (3) and $t \in T$. The definition of γ_j, $j \in T$, yields

(5) $$\gamma_j^{-1}(R) = \bigcap_{k=1}^{n} W_k^j \cap Y_j \quad \text{where}$$

(6) $$W_k^j := B_k^j \times \prod_{J \in A_j(\theta) \setminus \{J_k \cup \{j\}\}} S_J; \; B_k^j := \varphi_{J_k, J_k \cup \{j\}}^{-1}(A_{J_k}) .$$

Now (10.17) and (2) show that

$$\int 1_R\, p_t^T \circ \xi_T \, d \sum_{s \in T} m_s = \sum_{s \in T} \int_{\gamma_s^{-1}(R)} p_t^T \circ \xi_T \circ \gamma_s \, d\eta_s$$

$$= \sum_{s \in T} \int_{\bigcap_{k=1}^{n} W_k^s \cap Y_s} p_t^T \circ \pi_s^T \, d\eta_s$$

$$= \sum_{s \in T} \int_{\left(\bigcap_{k=1}^{n} \varphi_{J_k \cup \{s\}, T}^{-1}(B_k^s) \times \prod_{I \in A_s(\Theta) \setminus \{I\}} S_I\right) \cap Y_s} p_t^T \circ \pi_s^T \, d\eta_s$$

$$= \sum_{s \in T} \int_{\bigcap_{k=1}^{n} \varphi_{J_k, T}^{-1}(A_{J_k})} p_t^T \cdot p_s^T \, d\eta_I$$

$$= \int_{\bigcap_{k=1}^{n} \varphi_{J_k, T}^{-1}(A_{J_k})} p_t^T \, d\eta_I \, ,$$

since $\sum_{s \in T} p_s^T = 1$ and

$$(\varphi_{J_k \cup \{s\}, T}^{-1}(B_k^s) \times \prod_{I \in A_s(\Theta) \setminus \{T\}} S_I) \cap Y_s$$

$$= (\varphi_{J_k, T}^{-1}(A_{J_k}) \times \prod_{I \in A_s(\Theta) \setminus \{T\}} S_I) \cap Y_s \; .$$

On the other hand

$$\int_R dm_t = \eta_t \left(\bigcap_{k=1}^{n} W_k^t \cap Y_t \right)$$

$$= \eta_t\left(\left(\bigcap_{k=1}^{n} \varphi_{J_k \cup \{t\}, T}^{-1}(B_k^t) \times \prod_{I \in A_s(\Theta) \setminus \{T\}} S_I\right) \cap Y_t\right)$$

$$= \int_{\bigcap_{k=1}^{n} \varphi_{J_k, T}^{-1}(A_{J_k})} p_t^T \, d\eta_I \, ,$$

showing the validity of (iv) for $T_1 = T_2$.

6. According to (7.2) the following result holds for σ-finite measures ν_s , $s \in T_2$:

$$\left(\frac{d\nu_t}{d \sum_{s \in T_1} \nu_s} \right)_{t \in T_1} = \varphi_{T_1,T_2}\left(\left(\frac{d\nu_t}{d \sum_{s \in T_2} \nu_s}\right)_{t \in T_2} \right)$$

$\sum_{s \in T_1} \nu_s$ a.e. □

(10.18) Remarks. (a) We note that the integral part of $\Psi(z)$ can be represented in the form

$$\int_S (f_z{}^I \circ \xi_I - \sum_{j \in I} z_j P_j{}^I \circ \xi_I) d \sum_{j \in I} m_j$$

where $(m_j)_j$ denotes the family of Lévy measures on S given in Lemma (10.16).

(b) Observe that the Lévy measure η_I of (10.12) may be computed from

$$\eta_I = \sum_{i \in I} \eta_i{}_{|(\pi_i{}^I)^{-1}}{}^{\pi_i{}^I}(C\{e_I\})$$

for $|I| =: k \geq 2$. Hence, for every $h \in C^2_{lok}(S_k, \mathbb{R}^k)$ the integral part of $(E_t)_I$ can be represented by the family of Lévy measures $(\eta_i)_{i \in I}$.

(c) In LeCam, 1974, Chap. 8, it is implicitely used that every infinitely divisible regular experiment without Gaussian part arises from generalized Poisson processes. A first proof was given by Milbrodt for finite parameter spaces. In the general case, however, the problem was open for some time. Now, a careful application of Lemma (10.16) allows to carry this proof over to arbitrary parameter sets (compare with Part III of this volume).

Finally, we shall give a statistical interpretation of our results. Suppose that $(E_t = (X_t, A_t, (P_{i,t})_{i \in \Theta}))$ is a continuous semigroup of regular experiments. We consider the likelihood process

$$\left(\left(\frac{dP_{j,t}}{d \sum_{k \in I} P_{k,t}}\right)_j \right)_{I \in A_i(\Theta)} \quad \text{w.r.t.} \quad (X_t, A_t, P_{i,t})$$

with values in Y_i. Then $\mu_{i,t}{}^{\tau_i}$ is the distribution of this process which is the "projective limit of convolution semigroups". In this context the generating functional is the "derivative" of the underlying semigroup.

We note that $(E_t)_t$ is a Gaussian semigroup iff all Lévy measures η_i

vanish. Then E_t is homogeneous and $\left(\log \frac{dP_{j,t}}{dP_{i_o,t}}\right)_{j \in \Theta}$ is a Gaussian process w.r.t. $P_{i_o,t}$ $(t > 0)$ having the covariance structure $(2c_{ij})_{i,j \in \Theta}$ and the mean $(-c_{ii})_{i \in \Theta}$. If (c_{ij}) vanishes then E_t is called a <u>Poisson experiment</u>. As pointed out in the finite case every regular infinitely divisible experiment can be decomposed as a product of a Gaussian experiment and a Poisson experiment. We note that arbitrary experiments can be partitioned into regular subexperiments (see Lemma (10.1)) and that every subexperiment allows a treatment according to § 10.

III. REPRESENTATION OF POISSON EXPERIMENTS

Hartmut Milbrodt

11. Generalized Poisson Processes

In this chapter we take up the theory of Poisson experiments presented in Sec. 3 of the first part. To define standard Poisson experiments, which are Poisson experiments in the sense of LeCam, 1974, we first have to consider generalized Poisson processes and settle the question of their existence. Then it will be shown that every Poisson experiment in the sense of Definition (3.6) is equivalent to a standard Poisson experiment and vice versa.

Let $\mathcal{R}$ be a semi-ring of subsets of a non-empty set Y and $\mathbb{N}_0$ the additive semi-group of non-negative integers equipped with the discrete topology.

(11.1) Definition. A stochastic process $(\Omega, \mathcal{A}, P, (X_R)_{R \in \mathcal{R}})$ is a *process of independent increments* if

(1) $X_{R_1}, \ldots, X_{R_n}$ are independent whenever $\{R_1, \ldots, R_n\} \subset \mathcal{R}$ are pairwise disjoint.

(2) $X_{R_1} + X_{R_2} = X_{R_1 \cup R_2}$ if $R_1 \cap R_2 = \emptyset$, $\{R_1, R_2, R_1 \cup R_2\} \subset \mathcal{R}$.

A $\mathbb{N}_0$-valued process of independent increments $(\Omega, \mathcal{A}, P, (X_R)_{R \in \mathcal{R}})$ is a *generalized Poisson process* if $\mathcal{L}(X_R|P)$ is a Poisson distribution for every $R \in \mathcal{R}$.

Observe that for every generalized Poisson process $(X_R)_{R \in \mathcal{R}}$ $\mu: R \longmapsto E(X_R)$ is a content on $\mathcal{R}$; μ is called the *intensity* of the process.

If $\mathcal{R} = \{[s,t): 0 \leq s \leq t\}$ and $(Y_t)_{t \geq 0}$ is a Poisson process in the usual sense with parameter, say, c then

$$X_{[s,t)} := Y_t - Y_s, \quad [s,t) \in \mathcal{R},$$

defines a generalized Poisson process with intensity $c \cdot \lambda^1|_{\mathcal{R}}$. Conversely, every generalized Poisson process $(X_R)_{R \subset \mathcal{R}}$ with intensity $c \cdot \lambda^1|_{\mathcal{R}}$ yields a Poisson process with parameter c via

$$Y_t := X_{[0,t)}\ , \quad t \geq 0\ ,$$

provided all paths of (Y_t) belong to the Skorohod space $D[0,\infty)$.

Returning to a general semi-ring $\mathcal{R}$ our first aim is to prove the following theorem which is stated without proof in LeCam, 1974, Sec. 8.

(11.2) Theorem. For every positive content μ on $\mathcal{R}$ there is a generalized Poisson process with intensity μ.

Proof. As a first step we define a suitable projective system $(\Omega_\alpha, A_\alpha; \pi_{\alpha\beta}; P_\alpha; I)$ to which the Consistency theorem of Bochner (1955, Theorem 5.1.1.) will be applied.

Let ν_R denote the Poisson distribution with expectation $\mu(R)$ $(R \in \mathcal{R})$ and $(I, \leq)$ be the directed set of finite disjoint subfamilies of $\mathcal{R}$, preordered by refinement. For $\alpha = \{H_{1,\alpha}, \ldots, H_{n_\alpha,\alpha}\} \in I$ let $\Omega_\alpha := \mathbb{N}_0^{\alpha}$, A_α the appertaining Borel-σ-field and $P_\alpha := \bigotimes_{j=1}^{n_\alpha} \nu_{H_{j,\alpha}}$; if $\beta = \{H_{1,\beta}, \ldots, H_{n_\beta,\beta}\} \in I$ is a refinement of α, define the connecting map $\pi_{\alpha\beta} : \Omega_\beta \longrightarrow \Omega_\alpha$ by

$$\pi_{\alpha\beta}((\omega_{\beta,H_{j,\beta}})_{1 \leq j \leq n_\beta}) := \Big(\sum_{H_{i,\beta} \subset H_{j,\alpha}} \omega_{\beta,H_{i,\beta}}\Big)_{1 \leq j \leq n_\alpha}.$$

Apparently, the $\pi_{\alpha\beta}$ are continuous and onto and satisfy

$$\pi_{\alpha\gamma} = \pi_{\alpha\beta} \circ \pi_{\beta\gamma}\ , \quad \alpha \leq \beta \leq \gamma \text{ elements of } I\ .$$

We have to show that

$$P_\alpha = \mathcal{L}(\pi_{\alpha\beta} / P_\beta), \quad \{\alpha \leq \beta\} \subset I\ .$$

Obviously, it is sufficient to treat the case that β contains one more element than α, i.e.

$$\beta = \{H_{1,\beta}, \ldots, H_{n_\beta,\beta}\} = \{H_{1,\alpha}, \ldots, H_{n_\alpha,\alpha}, H_{n_\alpha+1,\beta}\} \quad \text{or}$$

$$\beta = \{H_{1,\beta}, \ldots, H_{n_\beta,\beta}\} = \{H_{1,\alpha}, \ldots, H_{n_\alpha-1,\alpha}, H_{n_\alpha,\beta}, H_{n_\alpha+1,\beta}\}$$

$$\text{where } H_{n_\alpha,\beta} \cup H_{n_\alpha+1,\beta} = H_{n_\alpha,\alpha}\ .$$

We restrict ourselves to the consideration of the second possibility. Let $B_j \subset \mathbb{N}_o$, $1 \le j \le n_\alpha$. Then

$$P_\beta(\pi_{\alpha\beta}^{-1}(\prod_{j=1}^{n_\alpha} B_j)) =$$

$$P_\beta\{\omega_\beta \in \Omega_\beta : \omega_{\beta,H_{j,\alpha}} \in B_j,\ 1 \le j \le n_\alpha - 1,\ \omega_{\beta,H_{n_\alpha,\beta}} + \omega_{\beta,H_{n_\alpha+1,\beta}} \in B_{n_\alpha}\} =$$

$$\prod_{j=1}^{n_\alpha-1} \nu_{H_{j,\alpha}}(B_j) \cdot (\nu_{H_{n_\alpha,\beta}} * \nu_{H_{n_\alpha+1,\beta}})(B_{n_\alpha}) =$$

$$\prod_{j=1}^{n_\alpha} \nu_{H_{j,\alpha}}(B_j) = P_\alpha(\prod_{j=1}^{n_\alpha} B_j)$$

which proves the compatibility.

Let $\Omega := \{(\omega_\alpha) \in \prod_{\alpha \in I} \Omega_\alpha : \pi_{\alpha\beta}(\omega_\beta) = \omega_\alpha,$ if $\alpha \le \beta\}$, the canonical projective limit of the system $(\Omega_\alpha, \pi_{\alpha\beta}, I)$. This system is "sequentially maximal" in the following sense: If $(\alpha_n) \subset I$ is an increasing sequence in I, and $\omega_n \in \Omega_{\alpha_n}$ are such that $\pi_{\alpha_n,\alpha_{n+1}}(\omega_{n+1}) = \omega_n$, $n \in \mathbb{N}$, then there is a $(\overline{\omega}_\alpha) \in \Omega$ such that $\overline{\omega}_{\alpha_n} = \omega_n$, $n \in \mathbb{N}$. To see this, note that all connecting maps $\pi_{\alpha\beta}$ are topologically perfect. Hence, considering the Alexandroff-compactifications $\Omega_\alpha \cup \{\infty_\alpha\}$ of Ω_α, $\alpha \in I$, $\overline{\pi}_{\alpha\beta}(\infty_\beta) := \infty_\alpha$ defines a continuous extension of $\pi_{\alpha\beta}$, $\{\alpha \le \beta\} \subset I$. By a well-known result on projective systems of compact spaces (cf. Engelking, 1977, Cor. 3.2.16) the "extended system" $(\Omega_\alpha \cup \{\infty_\alpha\}, \overline{\pi}_{\alpha\beta}, I)$ is sequentially maximal. Apparently, this gives sequential maximality of the original system.

Denote by A the σ-field on Ω generated by the projection maps $pr_\alpha : \Omega \longrightarrow \Omega_\alpha$, $\alpha \in I$. Then the Theorem of Bochner (loc. cit.) gives a probability measure P on A satisfying

$$L(pr_\alpha / P) = P_\alpha, \quad \alpha \in I.$$

Choose $X_R := pr_{\{R\}}$, $R \in R$. Then $\nu_R = P_{\{R\}} = L(X_R / P)$. To check that every trajectory of (X_R) is a $\mathbb{N}_o$-valued content on R is similarly simple. Finally, to verify the independence condition, let $\alpha = \{R_1,\dots,R_n\} \in I$. Then

$$L((X_{R_j})_{1 \le j \le n} / P) = P_\alpha = \bigotimes_{j=1}^{n} L(X_{R_j} / P). \qquad \square$$

<u>(11.3) Remark.</u> Keep the notation of the above proof. Similar to the theory of point processes a simpler process can be obtained. For this,

let Ω_0 be the set of $\mathbb{N}_0$-valued contents on $\mathcal{R}$, $A_0 \subset 2^{\Omega_0}$ the initial σ-field of the "evaluation maps" $Z_R: \omega \longmapsto \omega(R)$, $\nu \in \Omega_0$ $(R \in \mathcal{R})$, and define $J: \Omega \longrightarrow \Omega_0$ by $J((\omega_\alpha)_{\alpha \in I})(R) := \omega_{\{R\}}$, $R \in \mathcal{R}$. Then J is an isomorphism of measurable spaces, and the process

$$(\Omega_0, A_0, L(J/P), (Z_R)_{R \in \mathcal{R}})$$

has the desired properties, too.

In the case of a finite field we shall make use of the particular representation of generalized Poisson processes described in the next example.

<u>(11.4) Example.</u> Assume that $\mathcal{R}$ is a finite field in Y and μ a finite measure on $\mathcal{R}$. Let Ω_0, A_0 and $(Z_R)_{R \in \mathcal{R}}$ be defined as in the above remark. Furthermore, let $\Phi: Y \longrightarrow \Omega_0$ send every $y \in Y$ to the corresponding Dirac measure. If we put

$$P := e^{-\mu(Y)} \sum_{n=0}^{\infty} \frac{L(\Phi \mid \mu)^n}{n!},$$

where the exponent n indicates the n-th convolution power on the measurable semi-group (Ω_0, A_0) (with pointwise addition), then $(\Omega_0, A_0, P, (Z_R)_{R \in \mathcal{R}})$ is a generalized Poisson process with intensity μ (see Kerstan et al., 1974, 1.4.16 and 1.2.3).

<u>(11.5) Remark.</u> Let $(\Omega, A, P, (X_R)_{R \in \mathcal{R}})$ be a process of independent increments. Then

(*) $$\nu_R := L(X_R \mid P), \quad R \in \mathcal{R},$$

defines a <u>convolution semi-group on $\mathcal{R}$</u>:

$$\nu_R * \nu_S = \nu_{R \cup S}, \quad \text{if } \{R,S,R \cup S\} \subset \mathcal{R}, \ R \cap S = \emptyset.$$

If μ is a positive content on $\mathcal{R}$ and ν_R the Poisson distribution with expectation $\mu(R)$, $R \in \mathcal{R}$, then by Theorem (11.2) there is a process of independent increments satisfying (*). If $\mathcal{R} = \{[s,t): 0 \leq s \leq t < \infty\}$, then the well-known existence-theorem for Markov processes given the transition probabilities yields processes of independent increments satisfying (*). However, if both $\mathcal{R}$ and the semi-group (ν_R) are arbitrary, we do not know how to obtain the existence of a process of independent increments such that (*) holds.

The situation is even more complicated if the phase space if different from $\mathbb{R}^1$ or $\mathbb{N}_0$. For an existence-result with the multiplicative semigroup $[0, \infty)$ as phase space see Chapter IV, (18.3), where the theory of experiments with independent increments is employed to obtain processes of independent multiplicative increments.

12. Standard Poisson Experiments

Let $\mathcal{R}$ be a ring of subsets of a non-empty set Y .

(12.1) Definition. Let μ be a content on $\mathcal{R}$. The Poisson measure with intensity μ is the distribution $\pi(\mu)|_{(2^{\mathbb{N}_0})^{\mathcal{R}}}$ of a generalized Poisson process with intensity μ .

Evidently, any Poisson measure is uniquely determined by its intensity. Let $T \neq \emptyset$.

(12.2) Definition. An experiment E for the parameter space T is a standard Poisson experiment, if it is pairwise imperfect and if there are intensities μ_t , $t \in T$, on some ring $\mathcal{S}$ such that

$$E = (\mathbb{N}_0^{\mathcal{S}}, (2^{\mathbb{N}_0})^{\mathcal{S}}, \{\pi(\mu_t): t \in T\}) .$$

Apart from the requirement of pairwise imperfectness this is precisely the notion of Poisson experiments introduced by LeCam, 1974. The following theorem is the main result of this chapter.

(12.3) Theorem. Every Poisson experiment is equivalent to a standard Poisson experiment and vice versa. In particular, every projective limit of standard Poisson experiments is equivalent to a standard Poisson experiment.

For the proof we need a series of auxiliary results. First we determine the Hellinger transforms of standard Poisson experiments by "direct computation for ... a finitely generated field and passages to the limit" (LeCam, 1974, p. 73).

For the remainder of this section let μ_t , $t \in T$, be finite contents on $\mathcal{R}$ and $E := (\mathbb{N}_0^{\mathcal{R}}, (2^{\mathbb{N}_0})^{\mathcal{R}}, \{\pi(\mu_t): t \in T\})$, the standard Poisson experiment with intensities μ_t , $t \in T$.

(12.4) Lemma. Let $\mathcal{R}$ be a finite field. Then E is equivalent to the compound Poisson experiment with intensities μ_t , $t \in T$.

Proof. Let F be the compound Poisson experiment with intensities $(\mu_t)_{t\in T}$, $R_1,\dots,R_\ell$ the atoms of $\mathcal{R}$, $\alpha\in A(T)$ and $z\in S_\alpha$. Utilizing Lemma (3.3) we get

$$H(F_\alpha)(z) = \exp\Big(\sum_{j=1}^{\ell}\prod_{t\in\alpha}\mu_t(R_j)^{z_t} - \sum_{t\in\alpha} z_t\,\mu_t(Y)\Big) =$$

$$\exp\Big(-\sum_{t\in\alpha} z_t\,\mu_t(Y)\Big)\cdot\sum_{n=0}^{\infty}\ \sum_{m_1+\dots+m_\ell=n}\ \prod_{j=1}^{\ell}\frac{\prod_{t\in\alpha}\mu_t(R_j)^{z_t m_j}}{m_j!}\,.$$

On the other hand, for $t\in T$, let $(\Omega_o, A_o, P_t, (Z_R)_{R\in\mathcal{R}})$ be the Poisson process with intensity μ_t constructed in Example (11.4) and P the counting measure on Ω_o. If $(m_1,\dots,m_\ell)\in\mathbb{N}_o^{\ell}$ and $\lambda_{m_1,\dots,m_\ell}\in\Omega_o$ is defined by $\lambda_{m_1,\dots,m_\ell}(R_j) = m_j$, $j=1,\dots,\ell$, then

$$P_t(\lambda_{m_1,\dots,m_\ell}) = \exp(-\mu_t(Y))\cdot\prod_{j=1}^{\ell}\frac{\mu_t(R_j)^{m_j}}{m_j!}\,,\quad t\in\alpha\,.$$

Inserting this into

$$H(E_\alpha)(z) = \int_{\Omega_o}\prod_{t\in\alpha}\Big(\frac{dP_t}{dP}\Big)^{z_t}\,dP$$

leads to the above-obtained expression. □

Let N denote the directed set of finite subrings of $\mathcal{R}$ partially ordered by inclusion and $E^{S} := (\mathbb{N}_o^{S}, (2^{\mathbb{N}_o^{S}}), \{\pi(\mu_t|_S): t\in T\})$, $S\in N$.

(12.5) Lemma. For every $\alpha\in A(T)$ the net of Hellinger transforms $(H(E_\alpha^{S}))_{S\in N}$ decreases as S runs along N .

Proof. Let $\alpha\in A(T)$, $z\in S_\alpha$ and $S_1\subset S_2$ be subrings of $\mathcal{R}$ with atoms $A_1^{i},\dots,A_{\ell_i}^{i}$, $i=1,2$, respectively. Then

$$H(E_\alpha^{S_i})(z) = \exp\Big(\sum_{j=1}^{\ell_i}\prod_{t\in\alpha}\mu_t(A_j^{i})^{z_t} - \sum_{t\in\alpha} z_t\,\mu_t\big(\bigcup_{S\in S_i} S\big)\Big),\quad i=1,2$$

by the proof of the preceding lemma. Obviously, it is sufficient to treat the case $\ell_2 = \ell_1 + 1$, i.e.

$$A_j^1 = A_j^2 \ , \ j = 1,\dots,\ell_1 - 1 \ , \ A_{\ell_1}^1 = A_{\ell_1}^2 \cup A_{\ell_1+1}^2 \quad \text{or}$$

$$A_j^1 = A_j^2 \ , \ j = 1,\dots,\ell_1 \quad , \ A_{\ell_1+1}^2 \cap A_j^1 = \emptyset \ .$$

In the first case the assertion follows, since by Hölder's inequality

$$\prod_{t\in\alpha} \mu_t(A_{\ell_1}^1)^{z_t} \geq \prod_{t\in\alpha} \mu_t(A_{\ell_1}^2)^{z_t} + \prod_{t\in\alpha} \mu_t(A_{\ell_1+1}^2)^{z_t} ,$$

and in the second case we use Jensen's inequality for the logarithm to obtain

$$\prod_{t\in\alpha} \mu_t(A_{\ell_1+1}^2)^{z_t} \leq \sum_{t\in\alpha} z_t \mu_t(A_{\ell_1+1}^2)$$

and therefrom the assertion. □

(12.6) Lemma. $(E^S)_{S\in N}$ converges to E , weakly.

Proof. Since E^S is the image of E under the natural projection $\mathbb{N}_o^R \longrightarrow \mathbb{N}_o^S$ $(S\in N)$, E is more informative than every E^S . Together with the preceding lemma this shows that $(E^S)_{S\in N}$ converges weakly to an experiment $F = (\Omega, A, \{P_t : t\in T\})$ less informative than E .

We shall prove that F is also more informative than E . For the decision theoretic background of the following the reader is referred to the first two Chapters of LeCam, 1974, the terminology is taken from Strasser, 1984 b , Chapter X.

Let (D,W) be a standard decision problem, i.e. $D\subset R^T$ compact and convex and $W = (W_t)_{t\in D} = Id_D$. Denote by $R(F,D)$ the set of decision functions for (F,D) and by $R_s(E,D)$ the set of simple decision functions for (E,D); here a decision function ρ is called simple if it is non-randomized and if $\omega \longmapsto \text{supp}\ \rho(\omega,\cdot)$ takes on only finitely many values in D . For $t\in T$ let

$$||W_t||_\infty := \sup_{x\in D} |W_t(x)| \quad \text{and} \quad W_t\,\rho\,Q := \int\int W_t(x)\ \rho\,(\cdot,dx)dQ$$

if ρ is a kernel and Q a measure such that the double integral makes sense. Now, fix $\varepsilon > 0$ and $\alpha\in A(T)$. Recall that the comparison of experiments can be based solely on the comparison for standard decision problems; moreover, for such problems the set of risk vectors with respect to $R_s(E,D)$ $\{(W_t\,\rho\,\pi(\mu_t))_{t\in T} : \rho\in R_s(E,D)\}$ is dense in the set of all risk vectors. Hence, it is sufficient to show that for every

$\rho \in R_s(E,D)$ there is a $\sigma \in R(F,D)$ such that

$$\sum_{t \in \alpha} W_t \sigma P_t \leq \sum_{t \in \alpha} W_t \rho \pi(\mu_t) + \varepsilon \sum_{t \in \alpha} ||W_t||_\infty .$$

Let $\rho \in R_s(E,D)$, $\{d_1,\ldots,d_k\} \subset D$ and $A_1,\ldots,A_k$ be a measurable partition of $\mathbb{N}_0^R$ such that $\rho = \sum_{i=1}^{k} 1_{A_i} \varepsilon_{d_i}$ where ε_{d_i} denotes the Dirac measure sitting at d_i. Then

$$\sum_{t \in \alpha} W_t \rho \pi(\mu_t) = \sum_{t \in \alpha} \sum_{i=1}^{k} W_t(d_i) \pi(\mu_t)(A_i).$$

Choose $R_i \in N$ and $B_i \in (2^{\mathbb{N}_0^{R_i}})$ satisfying

$$\pi(\mu_t)(A_i \Delta (B_i \times \mathbb{N}_0^{R \setminus R_i})) < \frac{\varepsilon}{2k(k+1)}, \quad t \in \alpha, \quad 1 \leq i \leq k$$

and let $R_i \subset S \in N$ be chosen such that the deficiency of E^S and F is less than $\frac{\varepsilon}{2}$. Put

$$C_i := (B_i \times \mathbb{N}_0^{S \setminus R_i}) \setminus \bigcup_{j=1}^{i-1} (B_j \times \mathbb{N}_0^{S \setminus R_j}), \; 1 \leq j \leq k,$$

$$C_{k+1} := \mathbb{N}_0^S \setminus \bigcup_{i=1}^{k} C_i$$

and let

$$d_{k+1} \in D \setminus \{d_1,\ldots,d_n\}.$$

Then $\bar{\rho} = \sum_{i=1}^{k+1} 1_{C_i} \varepsilon_{d_i}$ is a decision function for E^S and

$$\left| \sum_{t \in \alpha} W_t \bar{\rho} \pi(\mu_t|_S) - \sum_{t \in \alpha} W_t \rho \pi(\mu_t) \right| \leq$$

$$\sum_{t \in \alpha} \sum_{i=1}^{k} |W_t(d_i)| \, |\pi(\mu_t)(C_i \times \mathbb{N}_0^{R \setminus S}) - \pi(\mu_t)(A_i)| +$$

$$\sum_{t \in \alpha} |W_t(d_{k+1})| \, \pi(\mu_t)(C_{k+1} \times \mathbb{N}_0^{R \setminus S}) \leq$$

$$\frac{\varepsilon}{2} \cdot \sum_{t \in \alpha} ||W_t||_\infty .$$

As we may find a $\sigma \in R(F,D)$ such that

$$\sum_{t \in \alpha} W_t \sigma P_t \leq \sum_{t \in \alpha} W_t \bar{\rho} \pi(\mu_t|_S) + \frac{\varepsilon}{2} \sum_{t \in \alpha} ||W_t||_\infty ,$$

the proof is complete. □

Combining the preceding lemmas we obtain

<u>(12.7) Corollary.</u> For every $S \in N$ let λ_S be a measure on $(\bigcup_{S \in S} S, S)$ dominating $\{\mu_t|_S: t \in T\}$. Then

$$H(E_\alpha)(z) = \exp\left(\inf_{S \in N} \int \psi_z\left(\left(\frac{d\mu_t|_S}{d\lambda_S}\right)_{t \in \alpha}\right) d\lambda_S\right), \quad z \in S_\alpha, \quad \alpha \in A(T).$$

Note that the pairwise imperfectness of E was irrelevant so far. The following fact serves for the simplification of the above obtained expression for the Hellinger transforms of E.

<u>(12.8) Remark.</u> Let $\alpha \in A(T)$ and μ_t, $t \in \alpha$, be restrictions of suitable measures λ_t, $t \in \alpha$, on the smallest σ-field $\sigma(R)$ containing R. Suppose there is a measure λ dominating $\{\lambda_t: t \in \alpha\}$ such that $\psi_z((\frac{d\lambda_t}{d\lambda})_{t \in \alpha})$ is λ-integrable for every $z \in S_\alpha$. Then

$$\int \psi_z\left(\left(\frac{d\lambda_t}{d\lambda}\right)\right) d\lambda = \inf_{S \in N} \int \psi_z\left(\left(\frac{d\mu_t|_S}{d\lambda|_S}\right)\right) d\lambda|_S, \quad z \in S_\alpha .$$

Since ψ_z is positively homogeneous, it suffices to sketch a proof for finite λ. Let $\varepsilon > 0$ and $z \in S_\alpha$. In view of the monotonicity shown in (12.5) we shall produce an $R_\varepsilon \in N$ such that

$$\left| \int \psi_z\left(\left(\frac{d\lambda_t}{d\lambda}\right)\right) d\lambda - \int \psi_z\left(\left(\frac{d\mu_t|_S}{d\lambda|_S}\right)\right) d\lambda|_S \right| < \varepsilon, \quad S \in N,\ R_\varepsilon \subset S .$$

For this, select an $R_0 \in R$ satisfying

$$\int_{X \setminus R_0} \left|\psi_z\left(\left(\frac{d\lambda_t}{d\lambda}\right)\right)\right| d\lambda < \frac{\varepsilon}{2} .$$

If $\lambda(R_0) = 0$, we are through. Otherwise, a standard martingale argument applied to

$$\frac{E_{\lambda|_{R_0 \cap \sigma(R)}}}{\lambda(R_o)}\left(\frac{d\mu_t|_{R_0 \cap \sigma(R)}}{d\lambda|_{R_0 \cap \sigma(R)}} \,\Big|\, S\right), \quad S \in N, \quad \bigcup_{R \in S} R = R_0$$

$(t \in \alpha)$ gives a subfield R_ε of the trace $R_0 \cap R$ of R on R_0 such that

$$\int_{R_0} \left| \frac{d\mu_t|_{R_0 \cap \sigma(R)}}{d\lambda|_{R_0 \cap \sigma(R)}} - \frac{d\mu_t|_S}{d\lambda|_S} \right| d\lambda \le \left(\frac{\varepsilon}{2|\alpha|}\right)^{\frac{1}{z_t}}, \quad t \in \alpha, \quad z_t > 0$$

for every subfield S of $R_o \cap R$ containing R_ε . It is a matter of mere routine to check that R_ε has the required property.

The proof that every standard Poisson experiment is in fact a Poisson experiment heavily relies on the convergence theory for Lévy measures.

(12.9) Lemma. Let T be finite and $(M_\nu)_{\nu \in N}$ a bounded net of Lévy measures on S_T such that

(1) $(\int \psi_z \, dM_\nu)_{\nu \in N}$ converges for every $z \in S_T$.

(2) $\lim\limits_{\varepsilon \to 0} \overline{\lim\limits_{\nu \in N}} \int_{\{s_T^2 < \varepsilon\}} s_T^2 \, dM_\nu = 0$.

Then there is exactly one Lévy measure M on S_T for which

$$\lim_{\nu \in N} \int \psi_z \, dM_\nu = \int \psi_z \, dM , \quad z \in S_T .$$

$(M_\nu)_{\nu \in N}$ converges to M , vaguely on $S_T \setminus \{e_T\}$.

Proof. Since $(M_\nu)_{\nu \in N}$ is bounded, we may find a Lévy convergent subnet $(M_{\nu_\lambda})_{\lambda \in L}$ of $(M_\nu)_{\nu \in N}$. Let (K,M) be the respective limiting pair (Part I, Lemma (4.6)). We show that K = 0 . For this, choose a decreasing sequence $f_n \in C(S_T)$, $n \in \mathbb{N}$, satisfying $1_{\{e_T\}} \leq f_n \leq 1_{\{s_T^2 < \frac{1}{n}\}}$, $n \in \mathbb{N}$. Then the proof of (4.6) (2), Part I , shows that

$$|K(s,t)| = |T| \left| \lim_{n \to \infty} \lim_{\lambda \in L} \int (p_s - \tfrac{1}{|\alpha|})(p_t - \tfrac{1}{|\alpha|}) f_n \, dM_{\nu_\lambda} \right|$$

$$\leq |T| \overline{\lim_{n \to \infty}} \, \overline{\lim_{\nu \in N}} \int_{\{s_T^2 < \frac{1}{n}\}} s_T^2 \, dM_\nu = 0 , \quad (s,t) \in T^2$$

by Assumption (2).

Now, take any convergent subnet of (M_ν) and let (K', M') denote its limiting pair. Then K' = 0 , and from Assumption (1) and Part I , Theorem (4.7) we obtain

$$\int \psi_z \, dM = \int \psi_z \, dM' , \quad z \in S_T ,$$

which implies M = M' by the Uniqueness theorem for Lévy measures. □

Partial proof of Theorem (12.3). Now, we are in a position to show that every standard Poisson experiment is a Poisson experiment.

W.l.g. assume that T is finite. Let M^S be the Lévy measure of E^S, $S \in N$. From Lemmas (12.5) and (12.6) we obtain

$$H(E)(z) = \exp\left(\lim_{S \in N} \int \psi_z \, dM^S\right), \quad z \in S_T .$$

Hence, for the existence of a Lévy measure of E it will be sufficient to prove that (M^S) fulfils the requirements of Lemma (12.9). By Lemmas (4.3), (12.6) and the pairwise imperfectness of E $(M^S)_{S \in N}$ is bounded. Let us establish Condition (12.9) (2).

If $R \in \mathcal{R}$ let G^R denote the standard Poisson experiment with intensities $\mu_t|_{R \cap \mathcal{R}}$, $t \in T$, where $R \cap \mathcal{R}$ denotes the trace of $\mathcal{R}$ on R. $(G^R)_{R \in \mathcal{R}}$ converges weakly to E, as can be checked by means of the Hellinger transforms. For every $\varepsilon > 0$, $R \in \mathcal{R}$ and every $S \in N$ we have

$$\int_{\{s_T^2 < \varepsilon\}} s_T^2 \, dM^S \leq \int_{S_T} s_T^2 \, d(M^S - M^{R \cap S}) + \int_{\{s_T^2 < \varepsilon\}} s_T^2 \, dM^{R \cap S}$$

where

$$\int_{\{s_T^2 < \varepsilon\}} s_T^2 \, dM^{R \cap S} \leq \varepsilon \, M^{R \cap S}(S_T) \leq \varepsilon \sum_{t \in T} \mu_t(R)$$

and

$$0 \leq \int s_T^2 \, d(M^S - M^{R \cap S}) \leq$$

$$\frac{4}{|T|} \sum_{s \in T} \sum_{t \in T} \int (\sqrt{p_s} - \sqrt{p_t})^2 \, d(M^S - M^{R \cap S}) \leq$$

$$\frac{8}{|T|} \sum_{s \in T} \sum_{t \in T} \Big[\left| \log H(E_{\{s,t\}})(\tfrac{1}{2}, \tfrac{1}{2}) - \log H(E^S_{\{s,t\}})(\tfrac{1}{2}, \tfrac{1}{2}) \right| +$$
$$\left| \log H(G^R_{\{s,t\}})(\tfrac{1}{2}, \tfrac{1}{2}) - \log H(E_{\{s,t\}})(\tfrac{1}{2}, \tfrac{1}{2}) \right| +$$
$$\left| \log H(G^R_{\{s,t\}})(\tfrac{1}{2}, \tfrac{1}{2}) - \log H(E^{R \cap S}_{\{s,t\}})(\tfrac{1}{2}, \tfrac{1}{2}) \right| \Big].$$

Taking into consideration the respective limit relations at our disposal, these inequalities show the validity of Condition (12.9) (2). □

For the proof of the converse we have to introduce some additional notation. Let

$$S := \prod_{\alpha \in A(T)} S_\alpha , \quad S^* := S \setminus \{(e_\alpha)_{\alpha \in A(T)}\} ,$$

$\pi^\alpha : S \longrightarrow S_\alpha$ the natural projection ,

$\mathcal{S}_\alpha := \sigma(\pi^\gamma : \gamma \subset \alpha)$ the σ-field generated by $(\pi^\gamma)_{\gamma \subset \alpha}$,

$$L_\alpha := S \setminus \bigcap_{\gamma \subset \alpha} (\pi^\gamma)^{-1}(e_\gamma) \quad (\alpha \in A(T)) \qquad \text{and}$$

$$R := \{A \in \bigotimes_{\alpha \in A(T)} B_\alpha : (e_\alpha)_\alpha \notin \bar{A}\} ,$$

the ring of relatively compact product measurable subsets of S^*. Furthermore, if $\{\alpha \subset \beta\} \subset A(T)$, let

$$N_{\alpha\beta} := \{z \in S_\beta : \sum_{t \in \alpha} z_t = 0\} \qquad \text{and}$$

$$\varphi_{\alpha\beta} : S_\beta \setminus N_{\alpha\beta} \ni (z_t) \longmapsto (\sum_{t \in \alpha} z_t)^{-1} (z_s) \in S_\alpha .$$

The following fundamental lemma is essentially due to Janssen (see Lemma (10.16) in the second chapter). For completeness we give a brief outline of its proof; details can be obtained from the proof of Lemma (10.16).

<u>(12.10) Lemma.</u> Let $(M_\alpha)_{\alpha \in A(T)}$ be a compatible system of Lévy measures, i.e.

$$\int_{S_\beta} \psi_z \, dM_\beta = \int_{S_\alpha} \psi_z \, dM_\alpha , \quad \{\alpha \subset \beta\} \subset A(T).$$

Then there is a family of premeasures m_t, $t \in T$, on the σ-ring $\sigma(R)$ generated by R such that

(1) $m_t(R) \subset [0, \infty)$, $t \in T$;

(2) $\int_{S^*} h \circ \pi^\alpha \, dm_t = \int_{S_\alpha} h \cdot p_t{}^\alpha \, dM_\alpha$, $h \in L^1(M_\alpha)$, $t \in \alpha \in A(T)$;

(3) for every $\{\alpha \subset \beta\} \subset A(T)$ $\varphi_{\alpha\beta} \circ \pi^\beta$ is defined $\sum_{s \in \alpha} m_s$ - a.e. and

$$\left(\frac{dm_t|_{S_\beta \cap L_\beta}}{d \sum_{s \in \alpha} m_s|_{S_\beta \cap L_\beta}} \right)_{t \in \alpha} = \varphi_{\alpha\beta} \circ \pi^\beta|_{L_\beta} .$$

For $t \in T$ let

$$A_t(T) := \{\alpha \in A(T) : t \in \alpha\}, \quad r(t) := (e_\alpha)_{\alpha \in A_t(T)} ,$$

$$Y_t := \{(z_\alpha) \in \prod_{\alpha \in A_t(T)} S_\alpha : \varphi_{\alpha\beta}(z_\beta) = z_\alpha , \ \{\alpha \subset \beta\} \subset A_t(T), \ z_\beta \notin N_{\alpha\beta}\}$$

the projective limit of $(S_\alpha, \varphi_{\alpha\beta}; A_t(T))$,

$$Y_t{}^* := Y_t \setminus \{r(t)\} \quad \text{and} \quad \pi_t{}^\alpha : Y_t \longrightarrow S_\alpha \ (\alpha \in A_t(T))$$

the canonical projection. As a first step towards the proof of Lemma (12.10) certain Radon measures η_t on Y_t^*, $t \in T$, will be constructed in Lemma (12.11) below. Next, defining $\gamma_t: Y_t^* \longrightarrow S^*$ by

$$\pi^\alpha \circ \gamma_t((z_\gamma)) := \begin{cases} z_\alpha & , \quad t \in \alpha \\ \varphi_{\alpha,\alpha \cup \{t\}}(z_{\alpha \cup \{t\}}) & , \quad t \notin \alpha , \ z_{\alpha \cup \{t\}} \notin N_{\alpha,\alpha \cup \{t\}} \\ e_\alpha & , \quad \text{otherwise} \end{cases}$$

$(\alpha \in A(T))$ it is not hard to check that γ_t is $\mathcal{B}(Y_t^*) - \underset{\alpha}{\otimes} \mathcal{B}(S_\alpha)$-measurable; hence $m_t := \mathcal{L}(\gamma_t | \eta_t)$ is well-defined $(t \in T)$. Following Lemma (12.11) we shall indicate how to verify properties (1) - (3) for (m_t).

<u>(12.11) Lemma.</u> (cf. Lemmas (10.13) and (10.17)) Let $(M_\alpha)_{\alpha \in A(T)}$ be a compatible system of Lévy measures. Then there is a family of Radon measures η_t on Y_t^*, $t \in T$, satisfying

(1) $$\int_{Y_t^*} h \circ \pi_t^\alpha \, d\eta_t = \int_{S_\alpha} h \cdot p_t^\alpha \, dM_\alpha , \quad t \in \alpha \in A(T) , \quad h \in L^1(M_\alpha) .$$

In particular,

(2) $$\eta_t((\pi_t^\beta)^{-1}(N_{\alpha\beta})) = 0 , \quad \{\alpha \subset \beta\} \subset A_t(T) ,$$

and:

(3) If $\{\alpha_1, \ldots, \alpha_n, \beta\} \subset A_t(T)$, $\bigcup_{k=1}^n \alpha_k \subset \beta$ and $A_k \in \mathcal{B}_{\alpha_k}$, $1 \leq k \leq n$, then
$$\eta_t\left(\left[\bigcap_{k=1}^n (\pi_t^{\alpha_k})^{-1}(A_k) \,\Delta\, (\pi_t^\beta)^{-1}\left(\bigcap_{k=1}^n \varphi_{\alpha_k\beta}^{-1}(A_k)\right)\right] \cap Y_t^*\right) = 0 \quad .$$

<u>Sketch of proof.</u> Recall that $C_b(X)$ denotes the space of real-valued bounded continuous functions on a topological space X ; $C_{oo}(X) \subset C_b(X)$ is the subspace of functions with compact support. Let $t \in T$ and define

$$T_t^\alpha := \{z \in S_\alpha : z_t > 0\} \quad (\alpha \in A_t(T)) \qquad \text{and}$$

$$V_t := \{f \in C_{oo}(Y_t^*) : f|_{(\pi_t^\alpha)^{-1}(T_t^\alpha)} = g \circ \pi_t^\alpha|_{(\pi_t^\alpha)^{-1}(T_t^\alpha)} \text{ for some } \alpha \in A_t(T) , \ g \in C_b(T_t^\alpha)\} .$$

Then

(i) V_t is a real algebra of functions which is positively rich in $C_{oo}(Y_t^*)$.

(ii) For every $f \in V_t$ there are an $\alpha \in A_t(T)$ and a $g \in C_b(T_t^\alpha)$ vanishing locally at e_α such that $f(x) = g \circ \pi_t^\alpha(x)$, $x \in (\pi_t^\alpha)^{-1}(T_t^\alpha)$.

Because of (i)' it suffices to define m_t on V_t. Let $f \in V_t$ and

$$f|_{(\pi_t^\alpha)^{-1}(T_t^\alpha)} = g \circ \pi_t^\alpha|_{(\pi_t^\alpha)^{-1}(T_t^\alpha)}$$

be a representation as in (ii). Put

$$\int_{Y_t^*} f \, d\eta_t := \int_{S_\alpha} g \, p_t^\alpha \, dM_\alpha .$$

Then $\int f \, d\eta_t$ is well-defined: If

$$f|_{(\pi_t^\beta)^{-1}(T_t^\beta)} = h \circ \pi_t^\beta|_{(\pi_t^\beta)^{-1}(T_t^\beta)}$$

is another representation as in (ii) and $\alpha \subset \beta$, then

$$h|_{T_t^\beta} = g \circ \varphi_{\alpha\beta}|_{T_t^\beta} .$$

Since

$$M_\alpha = L(\varphi_{\alpha\beta} | \sum_{s \in \alpha} p_s^\beta M_\beta)$$

by the Uniqueness theorem for Lévy measures (Theorem (3.8), Part I), it follows that

$$\int_{S_\beta} h \, p_t^\beta \, dM_\beta = \int_{S_\beta} g \circ \varphi_{\alpha\beta} \cdot p_t^\alpha \circ \varphi_{\alpha\beta} \, (\sum_{s \in \alpha} p_s^\beta) \, dM_\beta$$
$$= \int_{S_\alpha} g \cdot p_t^\alpha \, dM_\alpha .$$

If $\alpha \in A_t(T)$ and $h \in C_{oo}(S_\alpha \setminus \{e_\alpha\})$, then $h \circ \pi_t^\alpha \, V_t$ is a representation as in (ii); thus

$$\int_{Y_t^*} h \circ \pi_t^\alpha \, d\eta_t = \int_{S_\alpha} h \cdot p_t^\alpha \, dM_\alpha$$

showing that

$$L(\pi_t^\alpha | \eta_t)|_{B(S_\alpha \setminus \{e_\alpha\})} = p_t^\alpha M_\alpha|_{B(S_\alpha \setminus \{e_\alpha\})}$$

and hence (1).

(2) is an immediate consequence of (1), and (3) follows from (2) and the inequalities

$$0 \leq 1_{A_k} \circ \pi_t^{\alpha_k} - 1_{A_k} \circ \varphi_{\alpha_k\beta} \circ \pi_t^{\beta} \cdot 1_{(\pi_t^{\beta})^{-1}(S_\beta \setminus N_{\alpha_k\beta})} \leq 1_{N_{\alpha_k\beta}} \circ \pi_t^{\beta} ,$$

$$1 \leq k \leq n . \quad \square$$

(12.12) Remark. Let $\beta \in A(T)$, $\mathcal{R}_\beta := \mathcal{R} \cap S_\beta$ and denote by $\widetilde{\mathcal{R}}_\beta$ the set of those $R \in \mathcal{R}_\beta$ which can be represented as

$$(*) \qquad R = A_1 \times \ldots \times A_n \times \prod_{\alpha \in A(T) \setminus \{\alpha_1,\ldots,\alpha_n\}} S_\alpha , \quad \cup \alpha_i \subset \beta ,$$

$$S_{\alpha_i} \setminus A_i \text{ a measurable neighbourhood of } e_{\alpha_i} , \quad 1 \leq i \leq n .$$

Then $\mathcal{R}_\beta$ is a ring, $\widetilde{\mathcal{R}}_\beta$ is stable under finite intersections and

$$\sigma(\mathcal{R}_\beta) = \sigma(\widetilde{\mathcal{R}}_\beta) = S_\beta \cap L_\beta .$$

Sketch of proof for Lemma (12.10). Choose (m_t) as above. Then properties (1) and (2) follow almost immediately from Lemma (12.11) (1). For the proof of (3) let $\{\alpha \subset \beta\} \subset A(T)$. By (12.11) (2) we have

$$m_s(\pi^\beta \in N_{\alpha\beta}) = \eta_s(\pi^\beta \circ \gamma_s \in N_{\alpha\beta}) = \eta_s(\pi^\beta \in N_{\alpha\beta}) = 0 , \quad s \in \alpha .$$

Hence, $\varphi_{\alpha\beta} \circ \pi^\beta$ is defined $\sum_{s \in \alpha} m_s$ - a.e. Since

$$\left(\frac{dm_t|_{S_\beta \cap L_\beta}}{d \sum_{s \in \alpha} m_s|_{S_\beta \cap L_\beta}}\right)_{t \in \alpha} = \varphi_{\alpha\beta} \circ \left(\frac{dm_t|_{S_\beta \cap L_\beta}}{d \sum_{s \in \beta} m_s|_{S_\beta \cap L_\beta}}\right)_{t \in \beta} ,$$

it suffices to consider the case $\alpha = \beta$. In view of Remark (12.12) we shall then show that

$$\int_R p_t^\beta \circ \pi^\beta \, d \sum_{s \in \beta} m_s = \int_R dm_t$$

for every $t \in \beta$ and every R with representation (*). From Lemma (12.11) (3) we obtain

$$\gamma_t^{-1}(R) = (\pi_t^{\beta})^{-1} \left(\bigcap_{k=1}^{n} \varphi_{\alpha_k\beta}^{-1}(A_k)\right) \cap Y_t^* \quad \eta_t \text{ - a.e.}$$

Thus, making repeatedly use of Lemma (12.11) (1),

$$\int_R p_t^\beta \circ \pi^\beta \, d \sum_{s \in \beta} m_s =$$

$$\sum_{s \in \beta} \int_{Y_s^*} (1_{\bigcap_{k=1}^{n} \varphi_{\alpha_k\beta}^{-1}(A_k)} \cdot p_t^{\beta}) \circ \pi_s^{\beta} \, d\eta_s =$$

$$\sum_{s \in \beta} \int_{S_\beta} 1_{\bigcap_{k=1}^{n} \varphi_{\alpha_k\beta}^{-1}(A_k)} \cdot p_t^{\beta} p_s^{\beta} \, dM_\beta =$$

$$\int_{S_\beta} 1_{\bigcap_{k=1}^{n} \varphi_{\alpha_k\beta}^{-1}(A_k)} \cdot p_t^{\beta} \, dM_\beta =$$

$$\int_{Y_t^*} 1_{\bigcap_{k=1}^{n} \varphi_{\gamma_k\beta}^{-1}(A_k)} \circ \pi_t^{\beta} \, d\eta_t =$$

$$\int_R dm_t \, . \qquad \square$$

Final part of the proof of Theorem (12.3). Let F be a Poisson experiment for the parameter space T with Lévy measures M_α, $\alpha \in A(T)$. Then (M_α) is compatible in the sense of Lemma (12.10). To prove that F is equivalent to a standard Poisson experiment choose $m_t|_{\sigma(R)}$, $t \in T$, as above. Denote $\tilde{R} := \bigcup_{\beta \in A(T)} R_\beta$, the subring of cylinder sets with finite-dimensional base. In view of Theorem (11.2) the Poisson measures $\pi(m_t|_{\tilde{R}})$, $t \in T$, do exist. Let

$$G := (\mathbb{N}_0^{\tilde{R}}, (2^{\mathbb{N}_0^{\tilde{R}}}), \{\pi(m_t|_{\tilde{R}}) : t \in T\})$$

the standard Poisson experiment with intensities $m_t|_{\tilde{R}}$, $t \in T$, and $\alpha \in A(T)$. We shall show that $F_\alpha \sim G_\alpha$. Applying Corollary (12.7) and Remarks (12.8) and (12.12) we obtain for every $z \in S_\alpha$

$$H(G_\alpha)(z) =$$

$$\exp\left(\inf_{\beta \supset \alpha} \inf_{\substack{S \subset R_\beta \\ |S| < \infty}} \int \psi_z\left(\left(\frac{dm_t|_S}{d \sum_{s \in \alpha} m_s|_S} \right)_{t \in \alpha} \right) d \sum_{s \in \alpha} m_s|_S \right) =$$

$$\exp\left(\inf_{\beta \supset \alpha} \int \psi_z\left(\left(\frac{dm_t|_{S_\beta \cap L_\beta}}{d \sum_{s \in \alpha} m_s|_{S_\beta \cap L_\beta}} \right)_{t \in \alpha} \right) d \sum_{s \in \alpha} m_s|_{S_\beta \cap L_\beta} \right) =$$

$$\exp\left(\inf_{\beta \supset \alpha} \int_{L_\beta} \psi_z(\varphi_{\alpha\beta} \circ \pi^{\beta}) \, d \sum_{s \in \alpha} m_s|_{S_\beta \cap L_\beta} \right);$$

here the last identity follows from Lemma (12.10) (3). Since

$$\int_{L_\beta} \psi_z(\varphi_{\alpha\beta} \circ \pi^\beta)\, d \sum_{s \in \alpha} m_s|_{S_\beta \cap L_\beta} =$$

$$\sum_{s \in \alpha} \int \psi_z \circ \varphi_{\alpha\beta} \circ \pi^\beta \, dm_s =$$

$$\sum_{s \in \alpha} \int \psi_z \circ \varphi_{\alpha\beta} \cdot p_s^\beta \, dM_\beta \quad , \quad z \in S_\alpha , \quad \beta \supset \alpha ,$$

by Part (2) of the same lemma, and

$$M_\alpha = L(\varphi_{\alpha\beta} \mid \sum_{s \in \alpha} p_s^\beta M_\beta) \quad , \quad \beta \supset \alpha$$

by the Uniqueness theorem for Lévy measures (Theorem (3.8), Chap. I), we finally arrive at

$$H(G_\alpha)(z) = \exp(\int \psi_z \, dM_\alpha) = H(F_\alpha)(z) \, , \quad z \in S_\alpha .$$ □

From the construction contained in the preceding proofs we obtain

(12.13) Corollary. For every standard Poisson experiment there is an equivalent standard Poisson experiment with σ-additive intensities.

IV. STATISTICAL EXPERIMENTS WITH INDEPENDENT INCREMENTS

Helmut Strasser

13. Preliminaries

Let $F = (\Omega, A, \{P_\theta : \theta \in \Theta\})$ be an experiment whose parameter space $\Theta \subset \mathbb{R}^1$ is an open interval. Suppose that F is dominated by a σ-finite measure $\nu|A$ and denote

$$f(\cdot,\theta) := \frac{dP_\theta}{d\nu}, \quad \theta \in \Theta .$$

Fix some $\theta \in \Theta$ and assume that there exists a sequence $\delta_n \downarrow 0$ such that the sequence of experiments

$$E_{n,\theta} = (\Omega^n, A^n, \{P^n_{\theta+\delta_n t} : \theta + \delta_n t \in \Theta\}), \quad n \in \mathbb{N},$$

converges weakly to a nontrivial and pairwise imperfect limit experiment $E \in E(\mathbb{R}^1)$.

Ibragimov and Has'minskii, 1972 and 1981, study cases where the densities $f(\cdot,\theta)$, $\theta \in \Theta$, have jumps along some smooth lines and are smooth elsewhere. The conditions of Ibragimov and Has'minskii have been generalized by Pflug, 1983 . The limit experiments which are obtained by these authors are of a particular nature. First, one notes that they do not contain any Gaussian component and therefore should be Poisson experiments in the sense of LeCam's theory. Moreover, the logarithms of the likelihood processes are compound Poisson processes whose jump measures may give positive mass to $-\infty$.

In the present part we introduce a concept of experiments with independent increments of which the limit experiments of Ibragimov and Has'minskii and Pflug are special cases. We give necessary and sufficient conditions for triangular arrays to have weak limits with independent increments. With the results of the following sections the limit experiments of Ibragimov and Has'minskii and of Pflug can be characterized by three properties:

(1) The limit is a Poisson experiment in the sense of LeCam's theory.

(2) The limit has independent increments in a sense explained below.

(3) The Lévy measures of the limit are bounded.

The third property is necessary and sufficient for the fact that the likelihood process is equivalent to a compound Poisson process. For all three properties there are criteria which are much easier to verify than the sufficient conditions collected by Ibragimov and Has'minskii and by Pflug. We illustrate this fact in Section 19 .

The possible limit experiments having properties (1) - (3) are still simpler if they satisfy some invariance properties. The invariance property called scale invariance which has been introduced by Strasser, 1984 a , is necessarily satisfied in situations as described at the beginning. Moreover, translation invariance should be valid as is indicated by a theorem of LeCam, 1973. We show that a translation invariant experiment with independent increments is necessarily scale invariant with exponent $p = 1$. This is exactly the case which happens for the limits obtained by Ibragimov and Has'minskii and by Pflug. On the other hand this fact shows that the concept of experiments with independent increments does not cover asymptotic problems where the densities may have singularities since in such cases exponents $p \neq 1$ are not the exception (cf. Example (16.12) and Ibragimov and Has'minskii, 1981, Chapter VI).

14. Experiments with Independent Increments

Let $T := \{0,1,2,\ldots,k\}$. An experiment $E = (\Omega, A, \{P_0,P_1,\ldots,P_k\})$ for the parameter space T is abbreviated by $E = (P_0,P_1,\ldots,P_k)$.

<u>(14.1) Definition.</u> An experiment $E = (P_0,P_1,\ldots,P_k)$ in $E(T)$ has <u>independent increments</u> if

$$(P_0,P_1,\ldots,P_k) \sim (P_0,P_1,\ldots,P_1) \otimes (P_1,P_1,P_2,\ldots,P_2) \otimes \ldots \otimes (P_{k-1},\ldots,P_{k-1},P_k) \quad .$$

<u>(14.2) Lemma.</u> Suppose that $E = (P_0,P_1,\ldots,P_k)$ has independent increments. Then the Hellinger transform satisfies

$$H(E)(z) = \prod_{j=1}^{k} H(P_{j-1},P_j)\Big(\sum_{i=0}^{j-1} z_i, \sum_{i=j}^{k} z_i\Big), \quad z \in S_T \, .$$

<u>Proof.</u> For $j = 1,2,\ldots,k$ define

$$Q_i^{(j)} = \begin{cases} P_{j-1} & \text{if } 0 \leq i \leq j-1 \, , \\ P_j & \text{if } j \leq i \leq k \quad , \end{cases}$$

and $E_j = (Q_0^{(j)}, Q_1^{(j)}, \ldots, Q_k^{(j)})$, $1 \leq j \leq k$. Then $E \sim \prod_{j=1}^{k} E_j$ which implies

$$H(E) = \prod_{j=1}^{k} H(E_j).$$

Now the assertion follows from noting that

$$H(E_j)(z) = H(P_{j-1},P_j)\Big(\sum_{i=0}^{j-1} z_i, \sum_{i=j}^{k} z_i\Big), \quad z \in S_T \, . \qquad \square$$

<u>(14.3) Lemma.</u> Let $E = (P_0,P_1,\ldots,P_k)$ be an experiment whose Hellinger transform can be decomposed as

$$H(E)(z) = \prod_{j=1}^{k} f_j\Big(\sum_{i=j}^{k} z_i\Big), \quad z \in S_T \, ,$$

where $f_j: [0,1] \longrightarrow [0,1]$, $1 \leq j \leq k$, are arbitrary functions. Then E has independent increments and

$$f_j(z) = H(P_{j-1}, P_j)(1-z,z), \quad 0 \leq z \leq 1 .$$

Proof. In view of the proof of Lemma (14.2) it is obviously sufficient to show that $f_j(z) = H(P_{j-1}, P_j)(1-z,z), \; 0 \leq z \leq 1$.

For this we note that

$$1 = H(E)(1,0,\ldots,0) = f_1(1) \prod_{j=2}^{k} f_j(0) ,$$

$$1 = H(E)(0,0,\ldots,1) = \prod_{j=1}^{k} f_j(1) \qquad .$$

This implies $f_j(0) = f_j(1) = 1, \; 2 \leq j \leq k, \; f_1(1) = 1$. Now, fix $j \in \{1,2,\ldots,k\}$ and $z \in [0,1]$. Define $r_j(z) \in S_T$ by

$$r_{ji}(z) = \begin{cases} 0 & 0 \leq i \leq j-2 , \\ 1-z & i = j-1 \quad , \\ z & i = j \quad , \\ 0 & j+1 \leq i \leq k . \end{cases}$$

Then for every $j = 1,2,\ldots,k$

$$H(P_{j-1}, P_j)(1-z,z) = H(E)(r_j(z)) =$$

$$= \prod_{i=1}^{j-1} f_i(1) f_j(z) \prod_{i=j+1}^{k} f_j(0) =$$

$$= f_j(z) .$$

□

The following technical lemma will be of importance later. If $E = (P_0,P_1,\ldots,P_k)$ is an experiment in $E(T)$ let $\sigma|B(S_T)$ be the standard measure of E and $\sigma_{j-1,j}|B(S_{\{0,1\}})$ the standard measure of $(P_{j-1}, P_j), \; 1 \leq j \leq k$.

(14.4) Lemma. If $E = (P_0,P_1,\ldots,P_k)$ has independent increments then for every $\varepsilon > 0$

$$\sigma\{|x_{j-1} - x_j| \geq \varepsilon, \; |x_{\ell-1} - x_\ell| \geq \varepsilon\} \leq$$

$$\leq (k+1) \cdot \sigma_{j-1,j}\{|x_0-x_1| \geq \varepsilon\} \, \sigma_{\ell-1,\ell}\{|x_0-x_1| \geq \varepsilon\} ,$$

whenever $0 < j < \ell \leq k$.

Proof. Keep the notation of the proof of Lemma (14.2). Since the assertion only depends on the equivalence classes of the experiments, we prove the assertion for $F = \bigotimes_{j=1}^{k} E_j$.

Let $\sigma|A$ be a σ-finite measure dominating $P_o, P_1, \dots, P_k$, and denote $f_i := \frac{dP_i}{d\nu}$, $0 \le i \le k$. Let $Q_i = \prod_{j=1}^{k} Q_i^{(j)}$, $0 \le i \le k$. Then $F = (\Omega^k, A^k, \{Q_i : 0 \le i \le k\})$. Denote $g_i = \frac{dQ_i}{d\nu^k}$, $0 \le i \le k$, and $R = \sum_{i=0}^{k} Q_i$. Let $\varepsilon > 0$. Then

$$\sigma\{|x_{j-1} - x_j| \ge \varepsilon, |x_{\ell-1} - x_\ell| \ge \varepsilon\} =$$

$$= R\{|g_{j-1} - g_j| \ge \varepsilon(\Sigma g_m), |g_{\ell-1} - g_\ell| \ge \varepsilon(\Sigma g_m)\}.$$

The special construction of F shows that

$$g_{j-1} - g_j = \prod_{i=1}^{j-2} f_i (f_{j-1} - f_j) \prod_{i=j}^{k} f_i,$$

and

$$g_{j-1} + g_j = \prod_{i=1}^{j-2} f_i (f_{j-1} + f_j) \prod_{i=j}^{k} f_i.$$

Together with

$$|g_{j-1} - g_j| \ge \varepsilon(\Sigma g_m) \ge \varepsilon(g_{j-1} + g_j)$$

this implies

$$\{|g_{j-1} - g_j| \ge \varepsilon(\Sigma g_m)\} \subset p_j^{-1}\{|f_{j-1} - f_j| \ge \varepsilon(f_{j-1} + f_j)\} \quad R\text{-a.e.}$$

where p_j denotes the projection of Ω^k onto the jth coordinate. We obtain that

$$\sigma\{|x_{j-1} - x_j| \ge \varepsilon, |x_{\ell-1} - x_\ell| \ge \varepsilon\} \le$$

$$\le \sum_{m=0}^{k} Q_m(p_j^{-1}\{|f_{j-1} - f_j| \ge \varepsilon(f_{j-1} + f_j)\} \cap p_\ell^{-1}\{|f_{\ell-1} - f_\ell| \ge \varepsilon(f_{\ell-1} + f_\ell)\})$$

$$= \sum_{m=0}^{k} Q_m^{(j)}\{|f_{j-1} - f_j| \ge \varepsilon(f_{j-1} + f_j)\} Q_m^{(\ell)}\{|f_{\ell-1} - f_\ell| \ge \varepsilon(f_{\ell-1} + f_\ell)\})$$

$$\le \sum_{m=0}^{k} (P_{j-1} + P_j)\{|\frac{dP_{j-1}}{d(P_{j-1} + P_j)} - \frac{dP_j}{d(P_{j-1} + P_j)}| \ge \varepsilon\}$$

$$\cdot (P_{\ell-1} + P_\ell)\{|\frac{dP_{\ell-1}}{d(P_{\ell-1} + P_\ell)} - \frac{dP_\ell}{d(P_{\ell-1} + P_\ell)}| \ge \varepsilon\}$$

$$= (k+1)\,\sigma_{j-1,j}\{|x_o - x_1| \ge \varepsilon\}\sigma_{\ell-1,\ell}\{|x_o - x_1| \ge \varepsilon\}. \qquad \square$$

Let $T \neq \emptyset$ be a linearly ordered parameter set.

<u>(14.5) Definition.</u> An experiment $E = (\Omega, A, \{P_t: t \in T\})$ has <u>independent increments</u> if every subexperiment $\{P_{t_0}, P_{t_1}, \ldots, P_{t_k}\}$ with $t_0 < t_1 < \ldots < t_k$ has independent increments.

To motivate the reader we give a typical example of an experiment with independent increments for $T = \mathbb{R}^1$.

<u>(14.6) Example.</u> Let $F_\sigma = (\mathbb{R}^1, \mathcal{B}(\mathbb{R}^1), \{P_t^{(\sigma)}: t \in \mathbb{R}^1\})$ denote the translation parameter experiment of one-sided exponential distributions with fixed variance $\sigma > 0$, i.e.

$$\frac{dP_t^{(\sigma)}}{d\lambda^1}(x) = \frac{1}{\sigma}\exp\left(-\frac{x-t}{\sigma}\right) \cdot 1_{[0,\infty)}(x-t), \quad (x,t) \in \mathbb{R}^2$$

and $F_{-\sigma}$ its reflection at the origin. Then $F_1 \otimes F_{-1}$ is the weak limit experiment in the situation described in Section 13, when the densities $f(\cdot,\theta)$ are rectangular, $\theta \in \mathbb{R}^1$. If $t_0 < t_1 < \ldots < t_k$, then the Hellinger transforms are

$$H(P_{t_0}^{(\sigma)}, \ldots, P_{t_k}^{(\sigma)})(z) = \exp\left[\frac{1}{\sigma}\left(\sum_{i=0}^{k} z_i t_i - \max_{\{i: z_i > 0\}} t_i\right)\right],$$

$$z \in S_{\{t_0, \ldots, t_k\}}.$$

The decomposition of Lemma (14.3) is possible with

$$f_j(z) = \begin{cases} \exp\left[\frac{1}{\sigma}(t_{j-1} - t_j)(1-z)\right], & 0 < z < 1 \\ 1, & z = 0 \text{ or } z = 1. \end{cases}$$

Therefore, F_σ has independent increments. Similarly, $F_{-\sigma}$ has independent increments. Since any finite direct product of experiments with independent increments again is an experiment with independent increments, $F_\sigma \otimes F_{-\sigma}$ has independent increments, too.

An experiment with independent increments is completely described by its binary subexperiments.

<u>(14.7) Lemma.</u> Let $E = (\Omega_1, A_1, \{P_t: t \in T\})$ and $F = (\Omega_2, A_2, \{Q_t: t \in T\})$ be two experiments with independent increments. Then

$$E \sim F \quad \text{iff} \quad (P_s, P_t) \sim (Q_s, Q_t) \quad \text{for all } s < t,\ s \in T,\ t \in T.$$

<u>Proof.</u> This is obvious from Definitions (14.1) and (14.5). □

15. Existence and Construction of Experiments with Independent Increments

Let $T \neq \emptyset$ be a linearly ordered set.

(15.1) Definition. An incremental semigroup on T (of binary experiments) is a system $E_{s,t}$, $s<t$, $s \in T$, $t \in T$, of binary experiments satisfying

$$E_{s,t} \otimes E_{t,u} \sim E_{s,u} \quad \text{if} \quad s<t<u.$$

(15.2) Lemma. Let $E = (\Omega, \mathcal{A}, \{P_t: t \in T\})$ be an experiment with independent increments. Then the system $E_{s,t} = (P_s, P_t)$, $s<t$, $s \in T$, $t \in T$, is an incremental semigroup.

Proof. Let $s<t<u$. Since (P_s,P_t,P_u) has independent increments it follows by definition that

$$(P_s,P_t,P_u) \sim (P_s,P_t,P_t) \otimes (P_t,P_t,P_u).$$

Now let $z \in [0,1]$. Then

$$H(P_s,P_u)(1-z,z) = H(P_s,P_t,P_u)(1-z,0,z) =$$

$$= H(P_s,P_t,P_t)(1-z,0,z) \cdot H(P_t,P_t,P_u)(1-z,0,z) =$$

$$= H(P_s,P_t)(1-z,z) \cdot H(P_t,P_u)(1-z,z).$$

This proves the assertion. □

The correspondence between incremental semigroups and experiments with independent increments is total.

(15.3) Theorem. For every incremental semigroup $E_{s,t}$, $s<t$, $s \in T$, $t \in T$, there exists an experiment $E = (\Omega, \mathcal{A}, \{P_t: t \in T\})$ with independent increments such that

$$E_{s,t} \sim (P_s,P_t) \quad \text{for} \quad s<t,\ s \in T,\ t \in T.$$

Proof. For every $\alpha = \{t_0,t_1, \dots, t_k\} \subset T$ such that $t_0<t_1<\dots<t_k$ let E_α be such that

$$H(E_\alpha)(z) = \prod_{j=1}^{k} H(E_{t_{j-1},t_j})\left(\sum_{i=0}^{j-1} z_i, \sum_{i=j}^{k} z_i\right), \quad z \in S_\alpha .$$

We have to show that the experiments E_α, $\alpha \in A(T)$, form a projective system of experiments. Then any projective limit is of the desired nature.

Let $\beta = \alpha \setminus \{t_\ell\}$ for some $\ell = 0,1,2,\ldots,k$. For $z = (z_0,z_1,\ldots,z_{k-1}) \in S_\beta$ we denote $r_j = \sum_{i=j}^{k-1} z_i$, $0 \le j \le k-1$. Then we have

$$H((E_\alpha)_\beta)(z) = H(E_\alpha)(z_0,z_1,\ldots,z_{\ell-1}, 0, z_\ell,\ldots,z_{k-1}) =$$

$$= \prod_{j=1}^{\ell-1} H(E_{t_{j-1},t_j})(1-r_j,r_j)\cdot H(E_{t_{\ell-1},t_\ell})(1-r_\ell,r_\ell)$$

$$\cdot\ H(E_{t_\ell,t_{\ell+1}})(1-r_\ell,r_\ell)\cdot \prod_{j=\ell+2}^{k} H(E_{t_{j-1},t_j})(1-r_{j-1},r_{j-1}).$$

Since $E_{t_{\ell-1},t_\ell} \otimes E_{t_\ell,t_{\ell+1}} \sim E_{t_{\ell-1},t_{\ell+1}}$ it follows that

$$(E_\alpha)_\beta \sim E_\beta$$

which proves the assertion. □

Thus, the analysis of experiments with independent increments is reduced to the study of incremental semigroups. However, a further reduction is possible.

Recall from Strasser, 1984 b, that $M([0,\infty))$ denotes the set of probability measures $\mu|\mathcal{B}([0,\infty))$ such that $\int x\,\mu(dx) \le 1$. A binary experiment (P,Q) is completely described by the likelihood measures $\mu = L(\frac{dQ}{dP} \mid P) \in M([0,\infty))$. We denote by " $*$ " the convolution of measures in $M([0,\infty))$ with respect to the multiplication in $[0,\infty)$.

(15.4) Definition. A convolution semigroup on T is a system of measures $\mu_{s,t} \in M([0,\infty))$, $s<t$, $s \in T$, $t \in T$, satisfying

$$\mu_{s,t} * \mu_{t,u} = \mu_{s,u} \quad \text{if} \quad s<t<u.$$

The following is clear:

(1) If $E_{s,t}$, $s<t$, $s \in T$, $t \in T$, is an incremental semigroup, then $\mu_{s,t} := L(\frac{dP_t}{dP_s} \mid P_s)$, $s<t$, $s \in T$, $t \in T$, is a convolution semigroup.

(2) For every convolution semigroup $\mu_{s,t}$, $s<t$, $s\in T$, $t\in T$, there exists an incremental semigroup $E_{s,t}$, $s<t$, $s\in T$, $t\in T$, such that $\mu_{s,t}$ is the likelihood measure of $E_{s,t}$, $s<t$, $s\in T$, $t\in T$.

(15.5) Examples.

(1) For $T = \mathbb{R}^1$ define $\mu_{s,t} := \mathcal{L}(\exp|\nu_{\frac{t-s}{2},t-s})$, $s<t$, $s\in\mathbb{R}^1$, $t\in\mathbb{R}^1$. Then $\mu_{s,t}$, $s<t$, $s\in\mathbb{R}^1$, $t\in\mathbb{R}^1$, is a convolution semigroup and therefore defines an experiment with independent increments.

(2) If $\rho\in M([0,\infty))$ and $\lambda\geq 0$, then define

$$\exp(\lambda(\rho-\varepsilon_1)) := e^{-\lambda}\sum_{k=0}^{\infty}\frac{\lambda^k}{k!}\rho^{*k}.$$

Let $a>0$ be such that $\lambda(\int x\,\rho(dx)-1)+\log a\leq 0$. Then

$$\tau_t := \exp(t\lambda(\rho-\varepsilon_1)) * \varepsilon_{at}$$

is in $M([0,\infty))$ for every $t>0$. The measures $\mu_{s,t} := \tau_{t-s}$, $s<t$, $s\in\mathbb{R}^1$, $t\in\mathbb{R}^1$, form a convolution semigroup and therefore define an experiment with independent increments.

A particular case is the experiment $E_1 \otimes E_{-1}$ considered in Example (14.6) where $\rho=\varepsilon_0$, $\lambda=1$ and $a=1$.

The convolution semigroups considered in Example (15.5) have an invariance property.

(15.6) Definition. Let $T=\mathbb{R}^1$. A convolution semigroup $\mu_{s,t}$, $s<t$, $s\in\mathbb{R}^1$, $t\in\mathbb{R}^1$, is translation invariant, if $\mu_{s,t}=\mu_{s+u,t+u}$ whenever $s<t$, $u\in\mathbb{R}^1$, $s\in\mathbb{R}^1$, $t\in\mathbb{R}^1$.

(15.7) Theorem. Let $T=\mathbb{R}^1$. An experiment $E\in\mathcal{E}(T)$ with independent increments is translation invariant iff the convolution semigroup of likelihood measures is translation invariant.

Proof. Translation invariance of the convolution semigroup is equivalent to translation invariance of the Hellinger transforms of all binary subexperiments. Now, the assertion follows from Lemma (14.2). □

16. Infinitely Divisible Experiments with Independent Increments

Assume that $T \neq \emptyset$ is a linearly ordered set.

(16.1) Theorem. Assume that T is a topological space such that every order interval is connected. Then every continuous experiment with independent increments is infinitely divisible.

Proof. It is clear from Lemma (14.3) that an experiment E with independent increments is infinitely divisible iff each binary subexperiment is infinitely divisible.

Suppose that $E = (\Omega, A, \{P_t: t \in T\})$ is a continuous experiment with independent increments. For $s < t$, $s \in T$, $t \in T$, let

$$a(s,t) := -\log H(P_s,P_t)(\tfrac{1}{2},\tfrac{1}{2}).$$

Then the semigroup property implies

$$a(s,t) + a(t,u) = a(s,u) \quad \text{if} \quad s < t < u.$$

Now, fix $s < t$ and denote $b := d^2(P_s,P_t)$. If $b = 1$ or $b = 0$ it is trivial that (P_s,P_t) is infinitely divisible. Therefore we assume w.l.g. that $0 < b < 1$.

From continuity of E and connectedness of the intervals it follows that for every $n \in \mathbb{N}$ we may find points

$$s = u_{n,0} < u_{n,1} < \dots < u_{n,n-1} < u_{n,n} = t$$

such that $a(u_{n,i-1}, u_{n,i}) = \frac{1}{n} b$, $1 \leq i \leq n$. For every $n \in \mathbb{N}$ we have

$$(P_s,P_t) \sim \bigotimes_{i=1}^{n} (P_{u_{n,i-1}}, P_{u_{n,i}}).$$

Since

$$\max_{1 \leq i \leq n} d^2(P_{u_{n,i-1}}, P_{u_{n,i}}) = 1 - e^{-\frac{b}{n}}, \quad n \in \mathbb{N},$$

and

$$\sum_{i=1}^{n} d^2(P_{u_{n,i-1}}, P_{u_{n,i}}) = n(1 - e^{-\frac{b}{n}}), \quad n \in \mathbb{N},$$

the binary experiment (P_s,P_t) is the weak limit of a triangular array which is infinitesimal and bounded. Hence, (P_s,P_t) is infinitely divisible. □

Every infinitely divisible experiment is the product of a Gaussian experiment and of a Poisson experiment (Part I, Theorem (5.11)). We continue our discussion by considering Gaussian and Poisson experiments separately. The Gaussian case is particularly simple.

(16.2) Theorem. Let $E \in \mathcal{E}(T)$ be a Gaussian experiment. Then the following assertions are equivalent:

(1) E has independent increments.

(2) The system of binary subexperiments (P_s, P_t), $s < t$, is an incremental semigroup.

(3) The system of likelihood measures $\mathcal{L}(\frac{dP_t}{dP_s} \mid P_s)$, $s < t$, is a convolution semigroup.

(4) The Hellinger distances satisfy

$$(1 - d^2(P_s,P_t)) \cdot (1 - d^2(P_t,P_u)) = (1 - d^2(P_s,P_u)) \quad \text{if} \quad s < t < u.$$

Proof. The implications (1) $\Longrightarrow$ (2) $\Longrightarrow$ (3) $\Longrightarrow$ (4) are valid for every experiment. It remains to show that (4) $\Longrightarrow$ (1). Suppose that $E = (\Omega, \mathcal{A}, \{P_t: t \in T\})$ is a Gaussian experiment satisfying condition (4). Since for Gaussian experiments

$$\int \left(\frac{dP_t}{dP_s}\right)^z dP_s = (1 - d^2(P_s,P_t))^{4z(1-z)}, \quad 0 \le z < 1,$$

it follows that conditions (3) and (2) are satisfied.

Let $F \in \mathcal{E}(T)$ be an experiment with independent increments whose incremental semigroup is equivalent to (P_s,P_t), $s < t$, $s \in T$, $t \in T$. Then F is a Gaussian experiment since it is infinitely divisible and each binary subexperiment is Gaussian. Thus, E and F are Gaussian experiments whose binary subexperiments are equivalent. This implies that $E \sim F$ and therefore E has independent increments. □

Now, we derive the general form of the covariance structures of Gaussian experiments with independent increments.

(16.3) Lemma. The kernel

$$K(s,t) := \frac{1}{2} (|f(s)| + |f(t)| - |f(s) - f(t)|), \quad s,t \in T,$$

is a covariance kernel of a Gaussian experiment standardized at $t_0 \in T$

iff $f: T \longrightarrow \mathbb{R}^1$ satisfies $f(t_0) = 0$.

Proof. The condition is necessary since $0 = K(t_0,t_0) = |f(t_0)|$. For the converse let $E_0 = (\Omega, \mathcal{A}, \{P_t: t \in \mathbb{R}^1\})$ be a Gaussian experiment with covariance

$$K_0(s,t) = \frac{1}{2}(|s| + |t| - |s-t|), \quad s,t \in \mathbb{R}^1 .$$

Every experiment with independent increments for the convolution semigroup of Example (15.5), (1), is of this type. Let $f: T \longrightarrow \mathbb{R}^1$ be such that $f(t_0) = 0$. Then $E = (\Omega, \mathcal{A}, \{P_{f(t)}: t \in T\})$ is a Gaussian experiment with the desired covariance structure. □

(16.4) Theorem. A Gaussian experiment $E \in \mathcal{E}(T)$ has independent increments iff the covariance kernel K standardized at $t_0 \in T$ is of the form

$$K(s,t) = \frac{1}{2}(|f(s)| + |f(t)| - |f(s) - f(t)|), \quad s,t \in T,$$

where $f: T \longrightarrow \mathbb{R}^1$ is increasing.

Proof. First we show that the condition is necessary. We denote

$$a(s,t) := -\log H(P_s,P_t)(\tfrac{1}{2},\tfrac{1}{2}), \quad s,t \in T.$$

Since E has independent increments, it follows that

$$a(s,t) + a(t,u) = a(s,u) \quad \text{if} \quad s < t < u.$$

We define

$$\frac{1}{8} f(t) := \begin{cases} a(t, t_0) & \text{if} \quad t_0 < t , \\ -a(t, t_0) & \text{if} \quad t < t_0 . \end{cases}$$

Obviously, f is increasing. If K is the covariance kernel standardized at $t_0 \in T$ then we know from Discussion (2.9) in Part I that

$$K(s,t) = 4(a(s,t_0) + a(t,t_0) - a(s,t)), \quad s,t \in T.$$

Since

$$a(s,t) = \begin{cases} a(t,t_0) - a(s,t_0) & \text{if} \quad t_0 < s < t , \\ a(t,t_0) + a(s,t_0) & \text{if} \quad s < t_0 < t , \\ -a(t,t_0) + a(s,t_0) & \text{if} \quad s < t < t_0 , \end{cases}$$

it follows that

$$K(s,t) = \frac{1}{2}(|f(s)| + |f(t)| - |f(s) - f(t)|), \quad s,t \in T.$$

Now we prove the converse. We need only show that the Hellinger distances derived from covariance structures with increasing $f: T \longrightarrow \mathbb{R}^1$ satisfy Condition (16.2), (4). We know that

$$-\log(1 - d^2(P_s,P_t)) = \frac{1}{4}\left(\frac{K(s,s)+K(t,t)}{2} - K(s,t)\right) =$$
$$= \frac{1}{4}|f(s) - f(t)|, \quad s,t \in T \quad .$$

This proves Condition (16.2), (4), by easy computations. □

(16.5) Remarks.

(1) The covariance kernel K standardized at $t_o \in T$ of a Gaussian experiment with independent increments can be given in the form

$$K(s,t) = \begin{cases} f(s \wedge t) & \text{if } s > t_o, \quad t > t_o, \\ 0 & \text{if } s < t_o, \quad t > t_o, \\ -f(s \vee t) & \text{if } s < t_o, \quad t < t_o, \end{cases}$$

where $f: T \longrightarrow \mathbb{R}^1$ is increasing and $f(t_o) = 0$.

(2) Let $T = \mathbb{R}^1$. A Gaussian experiment $E \in \mathcal{E}(\mathbb{R}^1)$ has independent increments and is translation invariant iff the covariance kernel standardized at $t_o = 0$ is

$$K(s,t) = \lambda(|s| + |t| - |s-t|), \quad s,t \in \mathbb{R}^1, \quad \lambda \geq 0.$$

The proof of Lemma (16.3) shows that every Gaussian experiment with independent increments is equivalent to a reparametrization of E.

(3) Let $T = \mathbb{R}^1$. If a Gaussian experiment with independent increments is translation invariant and continuous, then it is scale invariant with exponent $p = 1$.

(4) Let $T = \mathbb{R}^1$. A Gaussian experiment with independent increments is scale invariant with exponent $p > 0$ iff it is continuous and the covariance kernel

$$K(s,t) = \frac{1}{2}(|f(s)| + |f(t)| - |f(s) - f(t)|), \quad s,t \in \mathbb{R}^1,$$

is such that

$$f(t) = \begin{cases} t^p f(1) & \text{if } t \geq 0, \\ |t|^p f(-1) & \text{if } t \leq 0. \end{cases}$$

Before we turn to Poisson experiments we establish a property of Lévy measures of experiments with independent increments. Throughout the following all experiments are supposed to be pairwise imperfect.

(16.6) Lemma. Let $E \in \mathcal{E}(T)$ be an infinitely divisible experiment with Lévy measures $(M_\alpha)_{\alpha \in A(T)}$. If E has independent increments then every Lévy measure M_α, $\alpha \in A(T)$, satisfies

$$M_\alpha\{x_s \neq x_t,\ x_u \neq x_v\} = 0 \quad \text{if} \quad s < t < u < v, \quad s,t,u,v \in \alpha .$$

Proof. Denote the n-th root of E by E_n and denote by $(\sigma_{n,\alpha})_{\alpha \in A(T)}$ the system of standard measures of E_n, $n \in \mathbb{N}$. It is clear that every E_n, $n \in \mathbb{N}$, has independent increments. From Theorem (5.9) and Corollary (4.8) in Part I we know that

$$\lim_{n\to\infty} n\,\sigma_{n,\alpha} = M_\alpha \quad \text{vaguely on} \quad S_\alpha \setminus \{e_\alpha\}, \quad \alpha \in A(T).$$

Letting $\alpha = \{t_0, t_1, \ldots, t_k\}$, $t_0 < t_1 < \ldots < t_k$, $k \in \mathbb{N}$, we need only show that for every $\varepsilon > 0$

$$\lim_{n\to\infty} n\,\sigma_{n,\alpha}\{|x_{j-1} - x_j| \geq \varepsilon,\ |x_{\ell-1} - x_\ell| \geq \varepsilon\} = 0$$

whenever $0 < j < \ell \leq k$. But this follows immediately from Lemma (14.4), since for every $\varepsilon > 0$

$$\overline{\lim_{n\to\infty}}\, n\,\sigma_{n,\{t_{j-1},t_j\}}\{|x_0 - x_1| \geq \varepsilon\} < \infty ,$$

and

$$\overline{\lim_{n\to\infty}}\, n\,\sigma_{n,\{t_{\ell-1},t_\ell\}}\{|x_0 - x_1| \geq \varepsilon\} < \infty . \qquad \square$$

The converse of Lemma (16.6) is not true in general, since every Gaussian experiment satisfies the condition. However, the following theorem is valid.

(16.7) Theorem. A Poisson experiment has independent increments iff every Lévy measure M_α, $\alpha \in A(T)$, satisfies

$$M_\alpha\{x_s \neq x_t,\ x_u \neq x_v\} = 0 \quad \text{if} \quad s < t < u < v, \quad s,t,u,v \in \alpha.$$

Proof. By Lemma (16.6) the condition is necessary. To prove sufficiency let $\alpha = \{t_0, t_1, \ldots, t_k\}$, $t_0 < t_1 < \ldots < t_k$, $k \in \mathbb{N}$. For every $j = 1,2, \ldots, k$ define $B_j \subset S_\alpha$ such that

$$B_j = \{x_0 = x_1 = \ldots = x_{j-1} \neq x_j = \ldots = x_k\}.$$

Then, by assumption we have $M_\alpha(S_\alpha \setminus \bigcup_{j=1}^{k} B_j) = 0$. It follows that for every $z \in S_\alpha$

$$\int (\prod_{t\in\alpha} x_t^{z_t} - \sum_{t\in\alpha} z_t x_t) M_\alpha(dx) =$$

$$= \sum_{j=1}^{k} \int_{B_j} (\prod_{t\in\alpha} x_t^{z_t} - \sum_{t\in\alpha} z_t x_t) M_\alpha(dx) =$$

$$= \sum_{j=1}^{k} \int_{B_j} (x_{t_{j-1}}^{1-r_j} x_{t_j}^{r_j} - (1-r_j)x_{t_{j-1}} - r_j x_{t_j}) M_\alpha(dx),$$

where $r_j = \sum_{i=j}^{k} z_i$, $1 \le j \le k$. Each term of the last sum is non-positive. Now, the assertion follows from Lemma (14.3). □

(16.8) Corollary. An infinitely divisible experiment has independent increments iff both, the Gaussian part and the Poisson part, have independent increments.

Proof. It is clear that the condition is sufficient. To prove necessity we first note that by Lemma (16.6) and Theorem (16.7) the Poisson part has independent increments. To show that also the Gaussian part has independent increments we establish Condition (16.2), (4).

Denote $E = (\Omega, A, \{P_t: t \in T\})$ and let $E \sim G \otimes H$ where $G = (\Omega_1, A_1, \{Q_t: t \in T\})$ is the Gaussian part and $H = (\Omega_2, A_2, \{R_t: t \in T\})$ is the Poisson part. It is clear that

$$(1 - d^2(P_s,P_t)) = (1 - d^2(Q_s,Q_t)) \cdot (1 - d^2(R_s,R_t)), \quad s,t \in T.$$

Since both E and H satisfy Condition (16.2), (4), it follows that also G satisfies Condition (16.2), (4). □

Let us study the incremental semigroups of Poisson experiments with independent increments. A binary Poisson experiment is completely described by its Lévy measure.

(16.9) Lemma. Let $E_{s,t}$, $s < t$, $s \in T$, $t \in T$, be an incremental semigroup of Poisson experiments whose Lévy measures are $M_{s,t}$, $s < t$, $s \in T$, $t \in T$. Then

$$M_{s,t} + M_{t,u} = M_{s,u} \quad \text{if} \quad s < t < u.$$

Proof. This is an immediate consequence of Remark (1) following Definition (15.4). □

(16.10) Remarks. Let $T = \mathbb{R}^1$.

(1) A Poisson experiment $E \in \mathcal{E}(T)$ with independent increments is translation invariant iff $M_{s,t} = (t-s)M_{0,1}$ whenever $s < t$. This is due to the fact that in case of translation invariance $M_{s,t} = M_{0,t-s}$, $s < t$, and $t \longrightarrow M_{0,t}(B)$ is additive and increasing for every $B \in \mathcal{B}(S_{\{0,1\}})$.

(2) A Poisson experiment which has independent increments and is translation invariant, is scale invariant with exponent $p = 1$.

(3) A Poisson experiment with independent increments is scale invariant with exponent $p > 0$ iff

$$M_{s,t} = \begin{cases} (t^p - s^p)M_{0,1} & \text{if} \quad 0 \leq s < t, \\ |s|^p M_{-1,0} + t^p M_{0,1} & \text{if} \quad s < 0 < t, \\ (|s|^p - |t|^p)M_{-1,0} & \text{if} \quad s < t \leq 0. \end{cases}$$

This follows from Lemma (16.9).

(16.11) Corollary. If a continuous infinitely divisible experiment has independent increments and is translation invariant then it is scale invariant with exponent $p = 1$.

Proof. Combine Corollary (16.8), Remarks (16.5), (3) and (16.10), (2). □

(16.12) Example. Let us take up Example (6.10) (a) of Part I, where the asymptotic behaviour of the localized one-sample translation parameter family of Γ-distributions with fixed variance $0 < \delta < 2$ has been analyzed. From the results obtained there it is clear that weak convergence takes place and that the limit experiment is scale invariant with exponent δ. Hence, by the above corollary, it cannot have independent increments if $\delta \neq 1$. For $\delta = 1$ it has independent increments in view of Example (14.6).

17. Weak Convergence of Triangular Arrays to Experiments with Independent Increments

Let $T \neq \emptyset$ be linearly ordered. We consider a triangular array of experiments $(E_{ni})_{1\le i\le k_n,\, n\in\mathbb{N}}$, denoting $E_{ni} = (\Omega_{ni}, A_{ni}, \{P_{nit}: t\in T\})$, $1 \le i \le k_n$, $n\in\mathbb{N}$. We are interested in the weak limit points of the product experiments $E_n = \prod_{i=1}^{k_n} E_{ni}$, $n\in\mathbb{N}$, which are denoted for short by $E_n = (\Omega_n, A_n, \{P_{nt}: t\in T\})$, $n\in\mathbb{N}$. Throughout the following we assume that the triangular array $(E_{ni})_{1\le i\le k_n,\, n\in\mathbb{N}}$ is bounded and infinitesimal. This implies that every weak limit is infinitely divisible and pairwise imperfect.

If $\alpha\in A(T)$ let $\sigma_{ni\alpha} = L\left(\left(\frac{dP_{nit}}{d\sum_{s\in\alpha} P_{nis}}\right)_{t\in\alpha} \Big| \sum_{s\in\alpha} P_{nis}\right)$, $1 \le i \le k_n$, $n\in\mathbb{N}$.

We denote $M_{n\alpha} := \sum_{i=1}^{k_n} \sigma_{ni\alpha}$, $\alpha\in A(T)$, $n\in\mathbb{N}$.

(17.1) Lemma. Suppose that $E_n \longrightarrow E$ weakly. Then the following assertions are equivalent:

(1) The Poisson part of E has independent increments.

(2) For every $\alpha\in A(T)$ and every $\varepsilon > 0$

$$\lim_{n\to\infty} M_{n\alpha}\{|x_s - x_t| \ge \varepsilon,\ |x_u - x_v| \ge \varepsilon\} = 0$$

whenever $s < t < u < v$, $s,t,u,v\in\alpha$.

Proof. Apply Theorem (16.7). □

Condition (2) of Lemma (17.1) can be simplified.

(17.2) Theorem. Suppose that $E_n \longrightarrow E$ weakly. Then the following assertions are equivalent:

(1) The Poisson part of E has independent increments.

(2') For every $r\in T$ and every $\varepsilon > 0$

$$\lim_{n\to\infty} \sum_{i=1}^{k_n} P_{nir}\{ \left|\frac{dP_{nit}}{dP_{nir}} - \frac{dP_{nis}}{dP_{nir}}\right| \geq \varepsilon, \left|\frac{dP_{niu}}{dP_{nir}} - \frac{dP_{niv}}{dP_{nir}}\right| \geq \varepsilon\} = 0$$

whenever $s < t < u < v$, $s,t,u,v \in T$.

<u>Proof.</u> (1) Assume that Condition (2') is satisfied. We prove that Condition (2) of Lemma (17.1) is satisfied. Let $s < t < u < v$, $s,t,u,v \in \alpha \in A(T)$. Then

$$M_{n\alpha}\{|x_s - x_t| \geq \varepsilon, |x_u - x_v| \geq \varepsilon\} =$$

$$= \sum_{i=1}^{k_n} \sum_{r\in\alpha} P_{nir}\{ \left|\frac{dP_{nis}}{dP_{nir}} - \frac{dP_{nit}}{dP_{nir}}\right| \geq \varepsilon \cdot \sum_{w\in\alpha} \frac{dP_{niw}}{dP_{nir}}, \left|\frac{dP_{niu}}{dP_{nir}} - \frac{dP_{niv}}{dP_{nir}}\right| \geq \varepsilon \cdot \sum_{w\in\alpha} \frac{dP_{niw}}{dP_{nir}}\} \leq$$

$$\leq \sum_{i=1}^{k_n} \sum_{r\in\alpha} P_{nir}\{ \left|\frac{dP_{nis}}{dP_{nir}} - \frac{dP_{nit}}{dP_{nir}}\right| \geq \varepsilon, \left|\frac{dP_{niu}}{dP_{nir}} - \frac{dP_{niv}}{dP_{nir}}\right| \geq \varepsilon\}$$

contains all relevant information.

(2) Conversely, assume that Condition (2) of Lemma (17.1) is satisfied. Let $s < t < u < v$, $s,t,u,v \in T$, and $r \in T$ arbitrary. Denote $R_{ni} = P_{nis} + P_{nit} + P_{niu} + P_{niv} + P_{nir}$. Then

$$\sum_{i=1}^{k_n} P_{nir}\{ \left|\frac{dP_{nit}}{dP_{nir}} - \frac{dP_{nis}}{dP_{nir}}\right| \geq \varepsilon, \left|\frac{dP_{niu}}{dP_{nir}} - \frac{dP_{niv}}{dP_{nir}}\right| \geq \varepsilon\} =$$

$$= \sum_{i=1}^{k_n} P_{nir}\{ \left|\frac{dP_{nit}}{dR_{ni}} - \frac{dP_{nis}}{dR_{ni}}\right| \geq \varepsilon \frac{dP_{nir}}{dR_{ni}}, \left|\frac{dP_{niu}}{dR_{ni}} - \frac{dP_{niv}}{dR_{ni}}\right| \geq \varepsilon \frac{dP_{nir}}{dR_{ni}}\} \leq$$

$$\leq \sum_{i=1}^{k_n} P_{nir}\{ \left|\frac{dP_{nit}}{dR_{ni}} - \frac{dP_{nis}}{dR_{ni}}\right| \geq \frac{\varepsilon}{a}, \left|\frac{dP_{niu}}{dR_{ni}} - \frac{dP_{niv}}{dR_{ni}}\right| \geq \frac{\varepsilon}{a}\} +$$

$$+ \sum_{i=1}^{k_n} P_{nir}\{\frac{dP_{nir}}{dR_{ni}} < \frac{1}{a}\}$$

for every $a > 0$. Thus, it remains to show that

$$\lim_{c\to\infty} \overline{\lim_{n\to\infty}} \sum_{i=1}^{k_n} P_{nir}\{\frac{dR_{ni}}{dP_{nir}} > c\} = 0 .$$

But this follows by standard arguments from the boundedness of the triangular array. □

(17.3) Corollary. Suppose that $E_n \longrightarrow E$ weakly. If $(E_{ni})_{1 \le i \le k_n}$, $n \in \mathbb{N}$ is a Poisson array, then E has independent increments iff (17.1), (2), or (17.2), (2'), is satisfied.

The Gaussian counterpart of this assertion is as follows.

(17.4) Theorem. Suppose that $E_n \longrightarrow E$ weakly. If $(E_{ni})_{1 \le i \le k_n}$, $n \in \mathbb{N}$ is a Gaussian array then E has independent increments iff

$$(3) \quad \lim_{n \to \infty} \sum_{i=1}^{k_n} (d^2(P_{nis}, P_{nit}) + d^2(P_{nit}, P_{niu}) - d^2(P_{nis}, P_{niu})) = 0$$

whenever $s < t < u$.

Proof. Since for every pair $s, t \in T$

$$\lim_{n \to \infty} \sum_{i=1}^{k_n} d^2(P_{nis}, P_{nit}) = - \log (1 - d^2(P_s, P_t)),$$

where $E = (\Omega, A, \{P_t: t \in T\})$, the condition is equivalent to (16.2), (4). □

(17.5) Corollary. Suppose that $E_n \longrightarrow E$ weakly. Then E has independent increments iff Conditions (2) and (3) or (2') and (3) are satisfied.

Proof. The proof is similar to the proof of Corollary (16.8). □

We want to dispense with the condition that $(E_n)_{n \in \mathbb{N}}$ converges weakly.

(17.6) Theorem. Suppose that $(E_{ni})_{1 \le i \le k_n}$, $n \in \mathbb{N}$ is a triangular array which satisfies (2) or (2'). Then $(E_n)_{n \in \mathbb{N}}$ converges weakly iff for some $t_o \in T$ and every $t \in T$ the sequences of binary experiments (P_{nt_o}, P_{nt}), $n \in \mathbb{N}$, converge weakly.

Proof. Let E and F be two weak limit points of $(E_n)_{n \in \mathbb{N}}$. Then (2) or (2') imply that E and F have Poisson parts with independent increments.

If

$$E = (\Omega_1, A_1, \{Q_t: t \in T\}),$$

and

$$F = (\Omega_2, A_2, \{R_t: t \in T\}),$$

then it follows that $(Q_{t_0}, Q_t) \sim (R_{t_0}, R_t)$ for every $t \in T$. In particular, the Poisson parts and the Gaussian parts of these binary experiments are equivalent. Hence, the Gaussian and the Poisson parts of E and F are equivalent. □

The main consequence of the preceding assertion is the fact that for a triangular array satisfying (2) or (2') the whole asymptotic behaviour is described by

$$L\left(\frac{dP_{n,t}}{dP_{n,t_0}} \mid P_{n,t_0}\right), \quad n \in \mathbb{N}, \quad t \in T,$$

for any fixed $t_0 \in T$. Thus, it can be handled in terms of the asymptotic theory of binary experiments.

<u>(17.7) Remark.</u> Let $T = \mathbb{R}^1$, $k_n = n$, $n \in \mathbb{N}$, and $E_{ni} = (\Omega, A, \{P_{\delta_n t}: t \in \mathbb{R}^1\})$, $1 \le i \le n$, where $\delta_n \downarrow 0$ satisfy (2) or (2'). If the sequence E_n, $n \in \mathbb{N}$, is equicontinuous, then every weak limit of $(E_n)_{n \in \mathbb{N}}$ is scale invariant and therefore the whole asymptotic behaviour is described by

$$L\left(\frac{dP_{\delta_n}}{dP_0} \mid P_0\right) \quad \text{and} \quad L\left(\frac{dP_{-\delta_n}}{dP_0} \mid P_0\right), \quad n \in \mathbb{N}.$$

Translation invariance is only possible if $(n\delta_n)_{n \in \mathbb{N}}$ varies slowly as $n \longrightarrow \infty$ (Strasser, 1984 a, Cor. (2.10)).

18. The Likelihood Process

Let $T \neq \emptyset$ be linearly ordered. We begin with a first basic property.

<u>(18.1) Theorem.</u> Suppose that $E = (\Omega, A, \{P_t: t \in T\})$ has independent increments. Let $t_o \in T$. Then the likelihood processes

$$\left(\frac{dP_t}{dP_{t_o}}\right)_{t < t_o} \quad \text{and} \quad \left(\frac{dP_t}{dP_{t_o}}\right)_{t > t_o}$$

are independent under P_{t_o}.

<u>Proof.</u> Let $s_\ell < s_{\ell-1} < \ldots < s_1 < t_o$ and $t_o < t_1 < \ldots < t_k$. We have to show that the random vectors

$$\left(\frac{dP_{s_i}}{dP_{t_o}}\right)_{1 \le i \le \ell} \quad \text{and} \quad \left(\frac{dP_{t_j}}{dP_{t_o}}\right)_{1 \le j \le k}$$

are independent under P_{t_o}. This will be done in the following way. We define two measures μ_1, μ_2 in $M([0,\infty)^{k+\ell})$. The first measure is

$$\mu_1 = L\left(\left(\frac{dP_{s_i}}{dP_{t_o}}\right)_{\ell \ge i \ge 1}, \left(\frac{dP_{t_j}}{dP_{t_o}}\right)_{1 \le j \le k} \,\Big|\, P_{t_o}\right).$$

For defining μ_2 we denote

$$f_{s_i}: (\omega_1,\omega_2) \longmapsto \frac{dP_{s_i}}{dP_{t_o}}(\omega_1), \quad (\omega_1,\omega_2) \in \Omega^2, \quad 1 \le i \le \ell,$$

$$g_{t_j}: (\omega_1,\omega_2) \longmapsto \frac{dP_{t_j}}{dP_{t_o}}(\omega_2), \quad (\omega_1,\omega_2) \in \Omega^2, \quad 1 \le j \le k.$$

Then $\mu_2 := L((f_{s_i})_{\ell \ge i \ge 1}, (g_{t_j})_{1 \le j \le k} \mid P_{t_o} \otimes P_{t_o})$.

The assertion is that $\mu_1 = \mu_2$. To prove this we compute the Mellin transforms $M(\mu_1)$ and $M(\mu_2)$. Let $z = (x_\ell, x_{\ell-1}, \ldots, x_1, y_1, \ldots, y_k)$ be such that $x_1 \ge 0$, $y_j \ge 0$, and $\sum_{i=1}^{\ell} x_i + \sum_{j=1}^{k} y_j < 1$. Denote $z_o := 1 - \sum_{i=1}^{\ell} x_i - \sum_{j=1}^{k} y_j > 0$. For convenience we identify $s_o := t_o$. Now,

$$M(\mu_1)(z) = H(P_{s_\ell}, P_{s_{\ell-1}}, \dots, P_{s_1}, P_{t_1}, \dots, P_{t_{k-1}}, P_{t_k})$$

$$(x_\ell, x_{\ell-1}, \dots, x_1, z_0, y_1, \dots, y_k) =$$

$$= H(P_{s_\ell}, P_{s_{\ell-1}})(x_\ell, 1-x_\ell) \cdot H(P_{s_{\ell-1}}, P_{s_{\ell-2}})(x_\ell + x_{\ell-1}, 1 - x_\ell - x_{\ell-1})$$

$$\dots \cdot H(P_{s_1}, P_{t_0})(x_1 + \dots + x_1, 1 - (x_\ell + \dots + x_1))$$

$$\cdot H(P_{t_0}, P_{t_1})(1 - (y_1 + \dots + y_k), y_1 + \dots + y_k) \cdot \dots$$

$$\dots \cdot H(P_{t_{k-1}}, P_{t_k})(1 - y_k, y_k).$$

On the other hand

$$M(\mu_2)(z) = \int \prod_{i=1}^{\ell} \left(\frac{dP_{s_i}}{dP_{t_0}}\right)^{x_i} dP_{t_0} \cdot \int \prod_{j=1}^{k} \left(\frac{dP_{t_j}}{dP_{t_0}}\right)^{y_j} dP_{t_0} =$$

$$= H(P_{s_\ell}, \dots, P_{s_1}, P_{t_0})(x_\ell, \dots, x_1, 1 - (x_1 + \dots + x_\ell)) \cdot$$

$$H(P_{t_0}, P_{t_1}, \dots, P_{t_k})(1 - (y_1 + \dots + y_k), y_1, \dots, y_k)$$

$$= H(P_{s_\ell}, P_{s_{\ell-1}})(x_\ell, 1 - x_\ell) \cdot \dots$$

$$\dots \cdot H(P_{s_1}, P_{t_0})(x_1 + \dots + x_\ell, 1 - (x_1 + \dots + x_\ell))$$

$$\cdot H(P_{t_0}, P_{t_1})(1 - (y_1 + \dots + y_k), y_1 + \dots + y_k) \cdot \dots$$

$$\dots \cdot H(P_{t_{k-1}}, P_{t_k})(1 - y_k, y_k). \qquad \square$$

In view of this theorem it is sufficient to consider the likelihood processes for $t < t_0$ and $t > t_0$ separately.

<u>(18.2) Definition.</u> A stochastic process $(\Omega, A, P; \{X_{s,t}: s<t, s \in T, t \in T\})$, realizing in $[0,\infty)$ is a <u>process of independent (multiplicative) increments</u> if

(1) X_{s_i,t_i}, $1 \le i \le k$, are independent whenever the intervals (s_i,t_i), $1 \le i \le k$, are pairwise disjoint,

(2) $X_{s,t} \cdot X_{t,u} = X_{s,u}$ if $s < t < u$.

(18.3) Remarks.

(1) If $(\Omega, A, P; \{X_{s,t}: s<t, s\in T, t\in T\})$ is a process of independent increments then the system of distributions $\mu_{s,t} = L(X_{s,t} \mid P)$, $s<t$, $s\in T$, $t\in T$, is a convolution semigroup.

(2) From Remark (2) below Definition (15.4) and Theorem (15.3) it follows that for every convolution semigroup $\mu_{s,t}$, $s<t$, $s\in T$, $t\in T$, there exists a process of independent increments $(\Omega, A, P; \{X_{s,t}: s<t, s\in T, t\in T\})$ such that $\mu_{s,t} = L(X_{s,t} \mid P)$, $s<t$, $s\in T$, $t\in T$.

(18.4) Theorem. Let $E = (\Omega_1, A_1, \{P_t: t\in T\})$ be an experiment with independent increments and $(\Omega_2, A_2, P; \{X_{s,t}: s<t, s\in T, t\in T\})$ a process of independent increments whose convolution semigroups coincide. Then for every $t_0 \in T$

$$L\left(\left(\frac{dP_t}{dP_{t_0}}\right)_{t>t_0} \Big| P_{t_0}\right) = L((X_{t_0,t})_{t>t_0} \mid P),$$

and

$$L\left(\left(\frac{dP_t}{dP_{t_0}}\right)_{t<t_0} \Big| P_{t_0}\right) = L((X_{t,t_0})_{t<t_0} \mid P).$$

Proof. We confine ourselves to the proof of the first assertion. Let $\mu_{s,t}$, $s<t$, $s\in T$, $t\in T$, be the underlying convolution semigroup and $\alpha = \{t_0, t_1,\ldots,t_k\}$, $t_0<t_1<\ldots<t_k$, $k\in \mathbb{N}$. Let $z\in S_\alpha$ such that $z_o>0$. Then

$$\int \prod_{j=1}^{k} \left(\frac{dP_{t_j}}{dP_{t_0}}\right)^{z_j} dP_{t_0} = H(P_{t_0}, P_{t_1},\ldots,P_{t_k})(z) =$$

$$= \prod_{j=1}^{k} H(P_{t_{j-1}}, P_{t_j})\left(\sum_{i=0}^{j-1} z_i, \sum_{i=j}^{k} z_i\right) =$$

$$= \prod_{j=1}^{k} M(\mu_{t_{j-1},t_j})\left(\sum_{i=j}^{k} z_i\right).$$

On the other hand we have

$$\int \prod_{j=1}^{k} (X_{t_0,t_j})^{z_j} dP = \int \prod_{j=1}^{k} \left(\prod_{i=1}^{j} X_{t_{i-1},t_i}\right)^{z_j} dP =$$

$$= \int \prod_{j=1}^{k} (X_{t_{j-1},t_j})^{\sum_{i=j}^{k} z_i} \, dP =$$

$$= \prod_{j=1}^{k} \int (X_{t_{j-1},t_j})^{\sum_{i=j}^{k} z_i} \, dP =$$

$$= \prod_{j=1}^{k} M(\mu_{t_{j-1},t_j})\left(\sum_{i=j}^{k} z_i\right) .$$

<u>(18.5) Discussion.</u> Let $T = \mathbb{R}^1$ and assume that $E = (\Omega, \mathcal{A}, \{P_t : t \in \mathbb{R}^1\})$ is a Poisson experiment with independent increments which is scale invariant with exponent $p > 0$. Assume further that the likelihood measures $\mu_{-1,0}$ and $\mu_{0,1}$ are bounded. This implies that

$$\mu_{0,1} = \exp(\lambda_1(\rho_1 - \varepsilon_1)) * \varepsilon_{a_1} ,$$

and

$$\mu_{-1,0} = \exp(\lambda_2(\rho_2 - \varepsilon_1)) * \varepsilon_{a_2} ,$$

where $\lambda_i \geq 0$, $\rho_i | \mathcal{B}([0,\infty))$ are probability measures and $a_i > 0$, $i = 1,2$, are such that

$$\lambda_i\left(\int x \, \rho_i(dx) - 1\right) + \log a_i \leq 0, \quad i = 1,2 .$$

For purposes of illustration we give an explicit construction of a process of independent increments for the convolution semigroup

$$\mu_{s,t} = \mathcal{L}\left(\frac{dP_t}{dP_s} \,\Big|\, P_s\right), \quad s < t, \quad s \in \mathbb{R}^1, \quad t \in \mathbb{R}^1 .$$

Let $(N_1(t))_{t > 0}$ and $(N_2(t))_{t > 0}$ be Poisson processes with intensities $(s,t) \longmapsto \lambda_i(t^p - s^p)$, $s < t$, $s \geq 0$, $t \geq 0$, $i = 1,2$, respectively. Let $(X_n)_{n \in \mathbb{N}}$ and $(Y_n)_{n \in \mathbb{N}}$ be sequences of i.i.d. random variables such $\mathcal{L}(X_n | P) = \rho_1$, $\mathcal{L}(Y_n | P) = \rho_2$, $n \in \mathbb{N}$. Choose things so that the processes $(N_1(t))_{t > 0}$, $(N_2(t))_{t > 0}$, $(X_n)_{n \in \mathbb{N}}$, $(Y_n)_{n \in \mathbb{N}}$ are mutually independent.

Then define

$$X_{s,t} := a_1^{|s|^p - |t|^p} \prod_{k=N_1(|t|)+1}^{N_1(|s|)} X_k \quad \text{if} \quad s < t \leq 0 ,$$

and

$$X_{s,t} := a_2^{t^p - s^p} \prod_{k=N_2(s)+1}^{N_2(t)} Y_k \quad \text{if} \quad 0 \leq s < t .$$

If $s < 0 < t$, then let $X_{s,t} := X_{s,0} \cdot X_{0,t}$. Easy computations show that $\{X_{s,t} : s < t,\ s \in \mathbb{R}^1,\ t \in \mathbb{R}^1\}$ is a process of the desired nature.

19. Application to Densities with Jumps

To illuminate the essentials we consider the general case of non-identically distributed observations. Let $k_n \uparrow \infty$ and consider σ-finite measure spaces $(\Omega_{ni}, \mathcal{A}_{ni}, \nu_{ni})$, $1 \leq i \leq k_n$, $n \in \mathbb{N}$. Let $T = \mathbb{R}^1$ and let $f_{nit}: \Omega_{ni} \longrightarrow \mathbb{R}^1$ be ν_{ni}-densities of probability measures $P_{nit} \mid \mathcal{A}_{ni}$, $t \in T$, $1 \leq i \leq k_n$, $n \in \mathbb{N}$. This defines a triangular array as considered in Section 17. We assume that the array is infinitesimal and bounded.

Now, we have to introduce conditions which describe the situation that the densities have isolated jumps. Conditions have been proposed by Ibragimov and Has'minskii, 1972 and 1981, and by Pflug, 1983, at least for the case of identically distributed observations. We propose the following conditions.

<u>(19.1) Conditions.</u> For every pair $s < t$ there are sets $M_{ni}(s,t) \in \mathcal{A}_{ni}$, $1 \leq i \leq k_n$, $n \in \mathbb{N}$, such that

(1) If $(s,t) \cap (u,v) = \emptyset$ then

$$M_{ni}(s,t) \cap M_{ni}(u,v) = \emptyset, \quad 1 \leq i \leq k_n, \quad n \in \mathbb{N}.$$

(2) $$\lim_{n\to\infty} \sum_{i=1}^{k_n} \int_{\Omega_{ni} \setminus M_{ni}(s,t)} (\sqrt{f_{nis}} - \sqrt{f_{nit}})^2 \, d\nu_{ni} = 0 .$$

(3) $$\overline{\lim_{n\to\infty}} \sum_{i=1}^{k_n} (P_{nis} + P_{nit})(M_{ni}(s,t)) < \infty .$$

Let us explain the ideas behind these conditions. If there are jumps, then the Hellinger distances are determined mainly by these jumps. On sets where jumps do not occur the contribution of the Hellinger distances is asymptotically negligible. Conditions (1) and (2) express the qualitative idea that the jumps can be included in disjoint sets. Condition (3) is of minor importance, more or less, as will turn out below.

<u>(19.2) Theorem.</u> Suppose that Conditions (19.1) (1) and (19.1) (2) are satisfied. Then the Poisson part of every weak limit of the triangular array has independent increments.

Proof. We have to show that Condition (2') of Theorem (17.2) is satisfied. Let $s < t < u < v$. Since $M'_{ni}(s,t) \cup M'_{ni}(u,v) = \Omega_{ni}$, $1 \leq i \leq k_n$, $n \in \mathbb{N}$, we obtain

$$\overline{\lim_{n\to\infty}} \sum_{i=1}^{k_n} P_{nir}\{|\frac{dP_{nis}}{dP_{nir}} - \frac{dP_{nit}}{dP_{nir}}| \geq \varepsilon, \; |\frac{dP_{niu}}{dP_{nir}} - \frac{dP_{niv}}{dP_{nir}}| \geq \varepsilon\} \leq$$

$$\leq \overline{\lim_{n\to\infty}} \sum_{i=1}^{k_n} P_{nir}\left(\{|\frac{dP_{nis}}{dP_{nir}} - \frac{dP_{nit}}{dP_{nir}}| \geq \varepsilon\} \cap M_{ni}(s,t)'\right) +$$

$$+ \overline{\lim_{n\to\infty}} \sum_{i=1}^{k_n} P_{nir}\left(\{|\frac{dP_{niu}}{dP_{nir}} - \frac{dP_{niv}}{dP_{nir}}| \geq \varepsilon\} \cap M_{ni}(u,v)'\right).$$

For the following we note that $|x-y| \geq \varepsilon$, $0 \leq x \leq a$, $0 \leq y \leq a$, implies

$$(\sqrt{x} - \sqrt{y})^2 = \frac{(x-y)^2}{(\sqrt{x} + \sqrt{y})^2} \geq \frac{\varepsilon^2}{4a}.$$

Hence, for every $a > 0$

$$\overline{\lim_{n\to\infty}} \sum_{i=1}^{k_n} P_{nir}\left(\{|\frac{dP_{nis}}{dP_{nir}} - \frac{dP_{nit}}{dP_{nir}}| \geq \varepsilon\} \cap M_{ni}(s,t)'\right) \leq$$

$$\leq \overline{\lim_{n\to\infty}} \frac{4a}{\varepsilon^2} \sum_{i=1}^{k_n} \int_{\Omega_{ni} \setminus M_{ni}(s,t)} \left(\sqrt{\frac{dP_{nis}}{dP_{nir}}} - \sqrt{\frac{dP_{nit}}{dP_{nir}}}\right)^2 dP_{nir} +$$

$$+ \overline{\lim_{n\to\infty}} \sum_{i=1}^{k_n} P_{nir}\{\frac{dP_{nis}}{dP_{nir}} \geq a\} + \overline{\lim_{n\to\infty}} \sum_{i=1}^{k_n} P_{nir}\{\frac{dP_{nit}}{dP_{nir}} \geq a\}.$$

Since

$$\lim_{a\to\infty} \overline{\lim_{n\to\infty}} \sum_{i=1}^{k_n} P_{nir}\{\frac{dP_{nis}}{dP_{nir}} \geq a\} = 0,$$

and similarly for the other term, the assertion follows. □

(19.3) Corollary. Suppose that Conditions (19.1) (1) - (19.1) (3) are satisfied. Then every weak limit of the triangular array is equivalent to a compound Poisson experiment.

Proof. First we prove that the limits are Poisson experiments. In view of Theorem (6.8), Part I, we have to show that for every pair $s,t \in \mathbb{R}^1$

$$\lim_{\varepsilon \to 0} \overline{\lim_{n \to \infty}} \sum_{i=1}^{k_n} \int_{\left| \frac{dP_{nit}}{dP_{nis}} - 1 \right| < \varepsilon} \left(\sqrt{\frac{dP_{nit}}{dP_{nis}}} - 1 \right)^2 dP_{nis} = 0 .$$

If $s < t$ this follows from

$$\sum_{i=1}^{k_n} \int_{\left| \frac{dP_{nit}}{dP_{nis}} - 1 \right| < \varepsilon} \left(\sqrt{\frac{dP_{nit}}{dP_{nis}}} - 1 \right)^2 dP_{nis} \leq$$

$$\leq \varepsilon \sum_{i=1}^{k_n} P_{nis}(M_{ni}(s,t)) +$$

$$+ \sum_{i=1}^{k_n} \int_{\Omega_{ni} \setminus M_{ni}(s,t)} \left(\sqrt{\frac{dP_{nit}}{dP_{nis}}} - 1 \right)^2 dP_{nis} .$$

If $s > t$ a similar argument is successful.

Secondly, we show that the Lévy measures of the limit experiment are bounded. It is sufficient to prove this for the Lévy measures $M_{s,t}$, $s < t$.

We have

$$M_{s,t}(S_{\{s,t\}}) = M_{s,t}\{x_s \neq x_t\} = \lim_{\varepsilon \to 0} M_{s,t}\{|x_s - x_t| \geq \varepsilon\} .$$

For a dense set of $\varepsilon \in (0,1)$

$$M_{s,t}\{|x_s - x_t| \geq \varepsilon\} =$$

$$\lim_{n \to \infty} \sum_{i=1}^{k_n} (P_{nis} + P_{nit}) \left\{ \left| \frac{dP_{nis}}{d(P_{nis} + P_{nit})} - \frac{dP_{nit}}{d(P_{nis} + P_{nit})} \right| \geq \varepsilon \right\}$$

$$\leq \overline{\lim_{n \to \infty}} \sum_{i=1}^{k_n} P_{nis} \left\{ \left| \frac{dP_{nit}}{dP_{nis}} - 1 \right| \geq \varepsilon \right\} +$$

$$+ \overline{\lim_{n \to \infty}} \sum_{i=1}^{k_n} P_{nit} \left\{ \left| \frac{dP_{nis}}{dP_{nit}} - 1 \right| \geq \varepsilon \right\} .$$

It follows that

$$M_{s,t}\{|x_s - x_t| \geq \varepsilon\} \leq \sum_{i=1}^{k_n} (P_{nis} + P_{nit})(M_{ni}(s,t)) < \infty$$

uniformly for $\varepsilon \in (0,1)$. □

Once having established properties of all possible limit experiments, it remains to prove actual weak convergence and to identify the limit. In view of Theorem (17.6) it is sufficient to consider the binary subexperiments. In the scale invariant case we need only consider the sequences $(P_{n,0}, P_{n,1})$, $n \in \mathbb{N}$, and $(P_{n,0}, P_{n,-1})$, $n \in \mathbb{N}$. They contain all necessary information on the likelihood processes as is shown in Discussion (18.5).

Finally, we show that our Conditions (19.1) are satisfied in situations which have been considered by Ibragimov and Has'minskii, 1981. The reader will note that we have to impose slightly stronger conditions than these authors. Indeed, we do not know whether the limit experiments obtained under their conditions in the weakest possible form (corresponding to $\delta = 0$ in Condition IV' below) are experiments with independent increments. Ibragimov and Has'minskii only establish the limit distribution of one particular likelihood process which does not determine the limit experiment. However, under slightly stronger conditions ($\delta > 0$) we will be able to show that the limits are in fact experiments with independent increments.

(19.4) Example. Ibragimov and Has'minskii, 1981, consider the case of identically distributed observations. We use their terminology (p. 242) and assume that their Conditions (I) - (III) are satisfied. Instead of their Condition (IV) we impose the slightly stronger condition

(IV'): For some $\delta \in (0,1]$ the integrals

$$\int \left| \frac{f'(\omega,\theta)}{f(\omega,\theta)} \right|^{1+\delta} f(\omega,\theta)\, \nu(d\omega) < \infty, \quad \theta \in \Theta,$$

and are continuous in $\theta \in \Theta$.

We show that assuming (I) - (III), (IV'), our Conditions (19.1) are satisfied.

We choose $M_n(s,t) := \bigcup_{k=1}^{r} [x_k(\theta + \frac{s}{n}), x_k(\theta + \frac{t}{n})]$, $s < t$. On the complements of these sets we have

$$\left(f(\omega,\theta + \tfrac{s}{n})^{1/2} - f(\omega,\theta + \tfrac{t}{n})^{1/2} \right)^2 \leq$$

$$\leq \left| f(\omega,\theta + \tfrac{s}{n})^{\frac{1}{1+\delta}} - f(\omega,\theta + \tfrac{t}{n})^{\frac{1}{1+\delta}} \right|^{1+\delta} =$$

$$= \left| \int_{\theta+\frac{s}{n}}^{\theta+\frac{t}{n}} \frac{f'(\omega,\theta+\xi)}{f(\omega,\theta+\xi)} f(\omega,\theta+\xi)^{\frac{1}{1+\delta}} d\xi \right|^{1+\delta} .$$

We obtain by Jensen's inequality

$$n \int_{\Omega\setminus M_n(s,t)} \left(f(\omega,\theta+\tfrac{s}{n})^{1/2} - f(\omega,\theta+\tfrac{t}{n})^{1/2} \right)^2 \nu(d\omega) \leq$$

$$\leq n \int_{\Omega} \left| \int_{\theta+\frac{s}{n}}^{\theta+\frac{t}{n}} \frac{f'(\omega,\theta+\xi)}{f(\omega,\theta+\xi)} f(\omega,\theta+\xi)^{\frac{1}{1+\delta}} d\xi \right|^{1+\delta} \nu(d\omega) =$$

$$= n \frac{(t-s)^{1+\delta}}{n^{1+\delta}} \int_{\Omega} \left| \frac{n}{t-s} \int_{\theta+\frac{s}{n}}^{\theta+\frac{t}{n}} \frac{f'(\omega,\theta+\xi)}{f(\omega,\theta+\xi)} f(\omega,\theta+\xi)^{\frac{1}{1+\delta}} d\xi \right|^{1+\delta} \nu(d\omega) \leq$$

$$\leq n \left(\frac{t-s}{n}\right)^{\delta} \int_{\Omega} \int_{\theta+\frac{s}{n}}^{\theta+\frac{t}{n}} \left| \frac{f'(\omega,\theta+\xi)}{f(\omega,\theta+\xi)} \right|^{1+\delta} f(\omega,\theta+\xi)\, d\xi \;\; \nu(d\omega)$$

$$= \left(\frac{t-s}{n}\right)^{\delta} n \int_{\theta+\frac{s}{n}}^{\theta+\frac{t}{n}} \int \left| \frac{f'(\omega,\theta+\xi)}{f'(\omega,\theta+\xi)} \right|^{1+\delta} f(\omega,\theta+\xi)\, \nu(d\omega)\, d\xi .$$

This implies that

$$\lim_{n\to\infty} n \int_{\Omega\setminus M_n(s,t)} \left(f(\omega,\theta+\tfrac{s}{n})^{1/2} - f(\omega,\theta+\tfrac{t}{n})^{1/2} \right)^2 \nu(d\omega) = 0$$

which proves (19.1), (2). Condition (19.1) (1) and (19.1) (3) follow immediately from Assumption (I) - (III).

(19.5) Example. Let us finish by applying our above results to the localized one-sample location parameter family of Pareto distributions with fixed scale parameter $\gamma > 0$ (cf. Example (6.10) (c) of Part I). It is not hard to check that the conditions of Example (19.4) are satisfied with any $\delta > 0$. Hence, every weak accumulation point of the respective sequence $(E_n)_n$ of product experiments is a Poisson experiment with bounded Lévy measures and independent increments. Moreover, defining $F_{\frac{1}{\gamma}}$ as in Example (14.6),

$$\lim_{n\to\infty} H(E_{n,\{s,t\}}) = H(F_{\frac{1}{\gamma},\{s,t\}}) \quad \text{on } S_{\{s,t\}}, \quad \{s,t\} \subset \mathbb{R}^1$$

which proves that $E_n \longrightarrow F_{\frac{1}{\gamma}}$, weakly, by Theorem (17.6).

BIBLIOGRAPHY

Becker, C. (1983). Schwache asymptotische Normalität von statistischen Experimenten bei unabhängigen, nicht notwendig identisch verteilten Beobachtungen. Dissertation, Bayreuther Mathematische Schriften 13, 1-153.

Blackwell, D. (1951). Comparison of experiments. Proc. 2nd Berkeley Symp. Math. Stat. Prob., 93-102.

Blackwell, D. (1953). Equivalent comparisons of experiments. Ann. Math. Statist. 24, 265-272.

Bochner, S. (1955). Harmonic Analysis and the Theory of Probability. University of California Press, Berkeley.

Bourbaki, N. (1965). Eléments de Mathématiques XIII, Livre VI: Intégration. Chapitre 3. Hermann, Paris, 2e ed.

Courrège, Ph. (1964). Générateur infinitésimal d'un semi-groupe de convolution sur $\mathbb{R}^n$ et formula de Lévy-Khintchine. Bull. Sci. Math., 2e Sér. 88, 3-30.

Engelking, R. (1977). General Topology. Pol. Scient. Publ., Warszawa.

Gnedenko, B.W. und A.N. Kolmogorov (1960). Grenzverteilungen von Summen unabhängiger Zufallsgrößen. Akademie-Verlag, Berlin.

Hájek, J. (1972). Local Asymptotic Minimax Admissibility in Estimation. Proc. 6th Berkeley Symp. Math. Stat. Prob., 175-194.

Hewitt, E. and K. Stromberg (1969). Real and Abstract Analysis. Springer, Berlin.

Heyer, H. (1977). Probability measures on locally compact groups. Springer, Berlin.

Ibragimov, I.A. and R.Z. Has'minskii (1972). The asymptotic behaviour of statistical estimates for samples with a discontinuous density. Math. USSR Sbornik 87 (129), 554-558.

Ibragimov, I.A. and R.Z. Has'minskii (1981). Statistical Estimation. Springer, Berlin.

Janssen, A. (1982). Unendlich teilbare statistische Experimente. Habilitationsschrift, Dortmund.

Kerstan, J.K., K. Matthes und J. Mecke (1974). Unbegrenzt teilbare Punktprozesse. Akademie-Verlag, Berlin.

Kruglov, N.M. (1970). A note on infinitely divisible distributions. Theory Prob. Appl. 15, 319-324.

Kruglov, N.M. (1974). On unboundedly divisible distributions in Hilbert spaces. Math. Notes Acad. Sci. USSR 16, 940-946.

LeCam, L. (1953). On some asymptotic properties of maximum likelihood estimates and related Bayes' estimates. Univ. of California, Publ. in Stat. 1, 277-330.

LeCam, L. (1960). Locally asymptotically normal families of distributions. Univ. of California, Publ. in Stat. 3, 37-98.

LeCam, L. (1964). Sufficiency and approximate sufficiency. Ann. Math. Statist. 35, 1419-1455.

LeCam, L. (1969). Théorie Asymptotique de la Décision statistique. Les Presses de l'Université de Montréal, Montréal.

LeCam, L. (1972). Limits of experiments. Proc. 6th Berkeley Symp. Math. Stat. Prob., Vol. 1, 245-261.

LeCam, L. (1974). Notes on asymptotic methods in statistical decision theory. I. Publ. du Centre de Recherches Mathématiques. Université de Montréal.

LeCam, L. (1979). On a theorem of J. Hájek. Contributions to Statistics. Hájek Memorial Volume. Editor: Dr. J. Jurečková; D. Reidel, Dordrecht; 119-135.

Millar, P.W. (1979). Asymptotic Minimax Theorems for the Sample Distribution Function. Z. Wahrscheinlichkeitstheorie verw. Geb. 48, 233-252.

Moussatat, M.W. (1976). On the Asymptotic Theory of Statistical Experiments and some of its Applications. Ph. D. Dissertation, Berkeley.

Oosterhoff, J. and W.R. van Zwet (1979). A Note on Contiguity and Hellinger Distance. Contributions to Statistics - Hájek Memorial Volume. Editor: Dr. J. Jurečková; D. Reidel, Dordrecht; 157-166.

Pfanzagl, J. (1980). Asymptotic expansions in parametric statistical theory. Developments in Statistics, Vol. 3, Editor: P.R. Krishnaiah; Academic Press, New York, 1-97.

Pflug, G. (1983). The Limiting Log-Likelihood Process for Discontinuous Density Families. Z. Wahrscheinlichkeitstheorie verw. Geb. 64, 15-35.

Prakasa Rao, B.L.S. (1968). Estimation of the location of the cusp of a continuous density. Ann. Math. Statist. 39, 76-87.

Siebert, E. (1979). Statistical Experiments and Their Conical Measures. Z. Wahrscheinlichkeitstheorie verw. Geb. 46, 247-258.

Siebert, E. (1982). Continuous convolution semigroups integrating a submultiplicative function. Manuscripta Math. 37, 383-391.

Strasser, H. (1984). Stability of Statistical Experiments. Accepted for publication in: Probability and Mathematical Statistics.

Strasser, H. (1984). Mathematical Theory of Statistics. Statistical experiments and asymptotic decision theory. De Gruyter, Berlin. (To appear)

Torgersen, E.N. (1970). Comparison of experiments when the parameter space is finite. Z. Wahrscheinlichkeitstheorie verw. Geb. 16, 219-249.

Torgersen, E.N. (1974). Asymptotic Behaviour of Powers of Dichotomies. Statistical Research Report No. 6, Oslo.

Torgersen, E.N. (1977). Mixtures and products of dominated experiments. Ann. Statist. 5, 44-64.

Wald, A. (1943). Tests of statistical hypotheses concerning several parameters when the number of observations is large. Trans. Amer. Math. Soc. 54, 426-482.

Waldenfels, W. von (1965). Fast positive Operatoren. Z. Wahrscheinlichkeitstheorie verw. Geb. 4, 159-174.

LIST OF SYMBOLS

$\mathcal{B}(X)$	Borel-σ-field of a topological space X
$M_b(X)$	set of bounded positive Radon measures on X
$M_1(X)$	$\subset M_b(X)$, subset of probability measures
$M[0,\infty)^k$	$\subset M_1[0,\infty)^k$, the set of probability measures with first moment bounded by 1
$C(X)$	$\{f: X \longrightarrow \mathbb{R}^1 : f \text{ continuous}\}$
$C_b(X)$	bounded functions in $C(X)$
$C_{oo}(X)$	$\subset C(X)$, continuous functions with compact support
supp f	support of $f \in C(X)$
$f\|_A$	a map f with domain A
$\overline{A}$	closure of A
$\overset{\circ}{A}$	interior of A
$\|A\|$	number of elements of A
CA, A'	complement of A
1_A	indicator function of A
δ_{ij}	$= 1_{\{i\}}(j)$, Kronecker's Delta
∇	Nabla Operator
h_x	partial derivative
$*$	convolution product of measures
P^{*n}	n-th convolution power of P
$\mathcal{L}(Y\|P)$, P^Y	distribution of Y w.r.t. P
$f\,\nu$	measure with ν-density f
$\frac{dP}{dQ}$	likelihood ratio of P w.r.t. Q
l.p.	log-likelihood process
$\nu \approx \mu$	the ν- and μ- null sets coincide
$\nu \perp \mu$	$\nu(A) = 0$ and $\mu(CA) = 0$ for some A
d	Hellinger distance of probability measures (cf. p. 16)
$\|\| \cdot \|\|$	variational norm of bounded signed measures

ε_x	Dirac (point) measure sitting at x
N_H	standard normal distribution on a Euclidean space H
$\mathcal{M}[0,\infty)^k$	$\subset M_1[0,\infty)^k$, the set of probability measures with first moment bounded by 1
λ^k	Lebesgue measure on the Euclidean k-space $\mathbb{R}^k$
φ	univariate standard normal Lebesgue density
Φ	univariate standard normal distribution function
$A(T)$	$\{\alpha \subset T:\ 1 \leq \lvert\alpha\rvert < \infty\}$, $T \neq \emptyset$
$A_t(T)$	$\{\alpha \in A(T):\ t \in \alpha\}$, $t \in T$
$p_t,\ p_t^T$	canonical projection of $[0,\infty)^T$ onto the t-th coordinate
$p_\alpha,\ p_\alpha^T$	canonical projection of $[0,\infty)^T$ onto $[0,\infty)^\alpha$
s_α^2	$\sum_{t\in\alpha} (p_t - \frac{1}{\lvert\alpha\rvert})^2$
S_α	$\{(z_t) \in [0,1]^\alpha:\ \sum_{t\in\alpha} z_t = 1\}$, the unit simplex in $\mathbb{R}^\alpha$
$\mathring{S}_\alpha$	$S_\alpha \cap (0,1)^\alpha$
$\mathcal{B}_\alpha$	$\mathcal{B}(S_\alpha)$
$N_{\alpha\beta}$	$\{(z_t) \in S_\beta:\ \sum_{t\in\alpha} z_t = 0\}$, $\{\alpha \subset \beta\} \subset A(T)$
$e_\alpha,\ e_k$	$(\frac{1}{\lvert\alpha\rvert}, \ldots, \frac{1}{\lvert\alpha\rvert}) \in S_\alpha$ $(\lvert\alpha\rvert = k)$
S	$\prod_{\alpha\in A(T)} S_\alpha$
$M(\mu)$	$z \longmapsto \int_{[0,\infty)^k} \prod_t x_t^{z_t}\, \mu(dx)$, $z \in S_\alpha$, the Mellin transform of $\mu \in \mathcal{M}([0,\infty)^k)$
$\mathcal{E}(\Theta)$	55
$\mathcal{E}I(\Theta)$	56
$\mathcal{E}IR(\Theta)$	84
D	84
Δ	14
$\sim$	15
$\dot{\mathcal{E}}$	55

E_α	15
$H(E)(z)$	16
H_ν	65
$\mathcal{T}_D$	84
$\mathcal{S}_D$	85
$E \otimes F$	16
$\overset{\infty}{\underset{n=1}{\oplus}} a_n E_n$	16
$\varepsilon(\mu)$	25
$E(\mu)$	59
$\pi(\mu)$	111
T_t^k, T_t^Θ	63, 64
G_k, G_Θ	63, 64
S^*	117
Y_t	91
$r(t)$	94
$\varphi_{I,J}$	56
φ_Θ, φ_k, φ_Θ^t	63, 92
$\mathcal{g}_t$, $\mathcal{g}$	72
τ_t	91
π_t^α	91
ψ_z	25
f_z	65
D_i, D_i^j	72, 92
$C^2_{lok}(\mathbb{R}^n)$	58
$C^2_{lok,A}(\mathbb{R}^n)$	60
$C^2_{lok}(T_i^k)$	72
$C^2_{lok}(S_k)$	72
$C^2_{lok}(S_k, \mathbb{R}^k)$	75
$\mathcal{D}(Y_t)$	92

AUTHOR INDEX

Becker, C.	44, 49, 50
Blackwell, D.	4, 55
Bourbaki, N.	97
Bochner, S.	13, 107, 108
Courrège, Ph.	58, 59
Gnedenko, B.W.	1
Hájek, J.	53
Has'minskii, R.Z.	2, 13, 43, 124, 149, 152
Heyer, H.	1, 58, 59
Ibragimov, I.A.	2, 13, 43, 124, 149, 152
Janssen, A.	67
Kakutani, N.	3
Kerstan, J.	109
Kolmogorov, A.N.	1
Kruglov	61
LeCam, L.	1, 3, 6, 7, 13, 15 ff., 21, 23, 28, 32, 37, 38, 45, 55, 80, 81, 83, 104, 106, 107, 111, 113, 125
Millar, P.W.	19
Moussatat, M.W.	19
Oosterhoff, J.	48
Pfanzagl, J.	6, 11
Pflug, G.	13, 21, 124, 149
Siebert, E.	31, 62
Strasser, H.	7, 9, 15, 21, 24, 44, 113, 125, 131
Torgersen, E.N.	55, 68
Wald, A.	1, 3, 6
Waldenfels, W. von	59, 87
Zwet, W.R. van	48

SUBJECT INDEX *

Almost positive functional 59, 84

Boundedness, of a net of Lévy measures 32
, of a triangular array of experiments 12, 32, 39, 42, 45, 47, 140

χ^2 - procedures 9
compatible Lévy measures 28, 118
conical measure 31
convolution semigroup (continuous) 58, 109, 131, 132, 146
, generating functional of a 58, 60, 84, 95
, translation invariant 132

Direct convex combination (of experiments) 16, 38
direct product (of experiments) 16, 38

Equivalence (class) of experiments 15
experiment (statistical) 2, 14
, binary 14, 83
, compound Poisson 25, 54, 69, 111
, Gaussian (shift) 4 ff., 19-24, 45, 87, 134-136
, homogeneous 16, 23, 71
, infinitely divisible 12, 38, 45, 56, 67 ff., 80 ff., 89, 133, 138
, pairwise imperfect 17, 26, 45, 68, 81, 111, 140
, Poisson 9 ff., 28, 30, 32, 36, 45, 88, 105, 111, 137-139, 150
, regular 68 ff., 81, 88
, scale invariant (= stable) 136, 139, 143, 147
, shift 2
, standard Poisson 13, 111, 123
, translation invariant 132, 136, 139
- types 15
, with independent increments 54, 126 ff.

* Underlined page numbers refer to definitions

exponential distributions	53, 129
Γ - distributions	43, 52
Gaussian array	<u>48</u>-50, 142
(shift) experiment	see experiment
measure (semigroup)	<u>60</u>
Hellinger distance	<u>16</u>
transform of a measure	<u>65</u> ff.
of an experiment	<u>16</u>
Infinitely divisible experiments	see experiments
probability measures	<u>58</u>
infinitesimality	12, <u>39</u>-42, 45, 47, 140
intensities of a compound Poisson experiment	<u>25</u>, 111
intensity, of a Poisson measure	<u>111</u>
, of a generalized Poisson process	<u>106</u>-109, 111
Kernel (of a Gaussian experiment)	20 ff.
, standardized at a point	<u>22</u>-24, 134, 136
s, equivalence of	<u>21</u>
LAN (Local Asymptotic Normality)	50, 52
Lévy convergence	<u>33</u>-36
-Khintchine formula/representation	
for convolution semigroups	58
for experiments	94
for Hellinger transforms	12, 46, 80, 98
for standard measures	75
measure	<u>28</u>, <u>59</u>, 99, 116, 118
, of compound Poisson semigroups	<u>69</u>
, of convolution semigroups	95
, of infinitely divisible experiments	<u>28</u> ff., 32-36, 47, <u>76</u> ff., <u>86</u>, 137, 138
likelihood (ratio) process	<u>15</u>, 144 ff.
measure	131
limit pair	34 ff., 116
localization	5, 43
Normalized exponential of a measure	<u>25</u>

Osterhoff-van Zwet criterion 48

Pareto distributions 44, 54, 153
Poisson array 50-53, 142
experiment see experiment
s accompanying a triangular array 45
measure 111
, compound 59
process, generalized 106, 107, 111
semigroup 59
process of independent (multiplicative) increments 106, 109, 145-147
projective (compatible) system of experiments 16, 31
limit of a projective system 16, 111

Regression problems 17, 41
restriction of an experiment 16

Semigroup, incremental 130, 131, 134, 138
semigroup of experiments 56, 67 ff., 72-75, 92, 94
, compound Poisson 69
, generating functional of a 75, 84
, Lévy measure of a 76
standard experiment 55 ff.
measure 55 ff., 67, 72, 75, 127

Tight functional 59
triangular array of experiments 17, 39 ff., 47 ff., 140-143, 149, 150
truncated normal distributions 43, 53

Weak convergence of experiments 7, 15

ERRATA

30_9 — $\partial_1 S_\alpha$ instead of $\partial_1 S$.

$30_{5,4}$ — Sufficiency only holds under additional assumptions (e.g. finiteness of the parameter set). Theorem (6.11) and Corollary (19.3) have to be adapted correspondingly.

36^{12-14} — S_α instead of S_T and K_j instead of K.

43^4 — Summation is from 1 to k (and not k_n).

$44^{7,9}$ — Replace $x+1$ by $x^{\gamma+1}$ and (4) by (1).

50^4 — (5.8) (a) instead of (5.8) (1).

51_7 — $\sum_{i=1}^{k_n}$ is missing before the $\int$.

$52^{3,4}$ —

$$\leq \int_{\bigcap_{q\in\alpha}\bigcap_{r\in\alpha}\{|\frac{dP_{niq}}{dP_{nir}}-1|<4\varepsilon|\alpha|\}} \sum_{u\in\alpha}\left(\frac{\frac{dP_{niu}}{dP_{nit}}}{\frac{dP_{nis}}{dP_{nit}}}-1\right)^2 \frac{dP_{nis}}{dP_{nit}}\, dP_{nit}$$

$$\leq \sum_{q\in\alpha}\int_{\{|\frac{dP_{niq}}{dP_{nis}}-1|<4\varepsilon|\alpha|\}}\left(\frac{dP_{niq}}{dP_{nis}}-1\right)^2 dP_{nis}.$$

52_3 — Replace dP_o by dQ_o.

93^4 — $D_j^i D_k^i$ instead of $D_j^i D_j^i$.

114_{10} — d_k instead of d_n.

120^{12} — Correct: $M_\alpha = 1_{S_\alpha \setminus \{e_\alpha\}} L(\varphi_{\alpha\beta} \mid \sum_{s\in\beta} p_s^\beta M_\beta)$.

The authors thank Dr. H. Zeuner, Tübingen, who pointed out these errors.

Lecture Notes in Statistics

Vol. 26: Robust and Nonlinear Time Series Analysis. Proceedings, 1983. Edited by J. Franke, W. Härdle and D. Martin. IX, 286 pages. 1984.

Vol. 27: A. Janssen, H. Milbrodt, H. Strasser, Infinitely Divisible Statistical Experiments. VI, 163 pages. 1985.